MW01671024

JMP IN® Companion for Robert Johnson's ELEMENTARY STATISTICS Seventh Edition

Al M. Best, III and R.K. Elswick, Jr.

Data Partners and
Virginia Commonwealth University

Duxbury Press
An Imprint of Wadsworth Publishing Company
I(T)P® An International Thomson Publishing Company

Belmont, CA • Albany, NY • Bonn • Boston • Cincinnati • Detroit • Johannesburg • London • Madrid
Melbourne • Mexico City • New York • Paris • San Francisco • Singapore • Tokyo • Toronto • Washington

Assistant Editor: Cynthia Mazow
Editorial Assistant: Martha O'Connor
Production: Al Best & R.K. Elswick/Data Partners
Print Buyer: Stacy Weinberger

Marketing Manager: Lauren Ward
Cover Designer: Stuart Paterson
Compositor: Data Partners

JMP IN® Companion for Robert Johnson's ELEMENTARY STATISTICS, Seventh Edition

Published by Duxbury Press, an imprint of Wadsworth Publishing Company, a division of International Thomson Publishing Inc. ITP The ITP logo is a registered trademark under license. Duxbury Press and the leaf logo are trademarks under license.

Printed in the United States of America
1 2 3 4 5 6 7 8 9 10

For more information, contact Wadsworth Publishing Company: 10 Davis Drive, Belmont, CA 94002, or electronically at http://www.thomson.com/wadsworth.html.

International Thomson Publishing Europe
Berkshire House 168-173
High Holborn
London, WC1V 7AA, England

Thomas Nelson Australia
102 Dodds Street
South Melbourne 3205
Victoria, Australia

Nelson Canada
1120 Birchmount Road
Scarborough, Ontario
Canada M1K 5G4

International Thomson Editores
Campos Eliseos 385, Piso 7
Col. Polanco
11560 México D.F. México

International Thomson Publishing GmbH
Königswinterer Strasse 418
53227 Bonn, Germany

International Thomson Publishing Asia
221 Henderson Road
#05-10 Henderson Building
Singapore 0315

International Thomson Publishing Japan
Hirakawacho Kyowa Building, 3F
2-2-1 Hirakawacho
Chiyoda-ku, Tokyo 102, Japan

International Thomson Publishing
Southern Africa
Building 18, Constantia Park
240 Old Pretoria Road
Halfway House, 1685 South Africa

ISBN 0-534-24317-7

Contents

Preface

The seventh edition of Robert Johnson's *Elementary Statistics* (ES7) is a modern text in many ways. For example, ES7 contains real-life data that are readily available for analysis. This is convenient because the best way to learn about data analysis and statistics is to work with data. Another feature of ES7 is its strong computer emphasis. Thankfully, gone are the days of unreachable hardware and difficult to use software. Modern software is easy to use and good software actually works with you to find out things in your data. We have been using JMP software since its beginning and are convinced that it is the best way to do data analysis and statistics for a wide range of common and not so common situations. So, it makes sense that this power and ease of use should be available to students. JMP IN is the student version of JMP and this companion is the student "how-to" manual whose aim is to show you how to do everything in Johnson's *Elementary Statistics*.

The driving thought behind this companion is, "If it's in the text, show how to do it with JMP IN." In particular, if you see a Minitab example in ES7, look in this companion to show you how to do it with JMP IN. All exercises use JMP IN 3.1.5 and do not include anything that is not in this student version of JMP.

The companion assumes the following:

- Students have ES7 and the student data disk.
- Students know how to work the Mac or Win operating system.
- Students have access either to the full JMP documentation or the JMP IN book *JMP Start Statistics*.

Not everything in ES7 is a simple series of JMP IN menu selections and option choices. Thus there are a number of programmed JMP templates documented in the Appendix. The JMP templates are available on disk.

We hope that you discover through JMP's use of modern graphics and interactive analysis that statistics and data analysis can be interesting and even exciting. Explore! Look for patterns and look for things that don't fit the pattern! Discover relations by fitting models! Don't be afraid to make mistakes and try out "dumb" ideas; it's easy to back off and try a different tack.

We welcome feedback. Al's eMail address is: Best@gems.VCU.edu and R.K.'s is: RKElswick@gems.VCU.edu. Our web homepages are: http://views.vcu.edu/~albest and http://views.vcu.edu/~rkelswick. When we think of anything of interest to JMP users or JMP companion readers, we post it there.

SAS Institute Inc.

JMP and JMP IN are from SAS Institute whose research headquarters is in Cary, North Carolina. You may want to visit their JMP homepage on the web: http://www.sas.com/jmp/. The Institute's flagship product is The SAS System, which is a much larger scale data processing package. The various versions of JMP are intended for people who want a nimble data analysis tool for a more intimate relationship with their data.

Data Partners

Data Partners is a private company primarily involved with JMP related consulting, training, and publishing. Al and RK are two of the company founders and have been involved with JMP since its first release.

Acknowledgments

A special thanks to Dan Chilko and Alex Kugushev for getting this project off the ground and seeing the value in a hands-on approach to the activity based teaching of statistics and data management.

It has also been a pleasure to work with the entire JMP team over the years. Keep the good stuff coming!

The reviewers for this companion were: Dan Chilko, Jenny Kendall, Ann Lehman, and Robert Johnson. We appreciate all of their advice a great deal. When you find mistakes, don't blame them.

Al sends a special thanks to Barbara, Ryan and Anna Kathryn. R.K. wishes to thank his best friend, Karen, and his kids: Trey, Amanda and Charlie.

Typographical Conventions

The typefaces used in this JMP companion were chosen to help you navigate the menus and options available.

- File names and data table names are capitalized. The files on the student disk that you will be using are the SAS transport files (EX2-010.XPT). When you import them from disk into the JMP's memory, they are termed data tables (EX2-010 on the Macintosh or EX2-010.JMP under Windows).
- The column names have a similar appearance to the way they appear in the JMP data table (WEIGHT).
- JMP is driven by menu and mouse commands. Menu names (**File** menu) or menu items (**Open** command) appear in a bold sans-serif font similar to the way they appear at the top of your screen. When you need to choose a command from the menu, the text appears as follows:
  Choose **File>Open**.
  This means that you should use the mouse to go to the **File** menu and choose the **Open** command. Some commands have keyboard equivalents that you can use instead.
- Similarly, JMP has a series of popup menu commands on the window border or inside the window being referred to. These commands appear as:
  Choose the **Analyze▸ Fit Line** option.
- Definitions and special notes appear in a shaded paragraph.

JMP?

JMP is pronounced "jump." In this text, JMP and JMP IN are used interchangeably. JMP IN is the student version of JMP distributed by Duxbury. Both JMP and JMP IN were developed by a small, extraordinarily talented group at SAS Institute headed by John Sall. John's Macintosh Product was designed from the beginning to be wholly interactive and easy to use. The Windows version works the same way that the Mac version does, with just a few exceptions.

- When there is a series of actions you should take to complete an example, the steps are shown as a list with gray bullets in the margin. For instance:

Whenever you need help on something in a JMP IN window, a context-sensitive interactive help system is just two clicks away.

- Choose **Tools>?**.
- Click on the item in the window.

The help window will stay on screen as long as your mouse is inside the help window. To return to the standard arrow tool, return to the **Tools** menu.

- Choose **Tools>**↖.

- JMP IN runs under either the Macintosh or Windows operating systems. Nearly all of the menus and options are identical, as are the contents of all of the windows. However, the window frames are different. To make this companion useful for all operating systems we only show the inside of a window, the "picture" but not the "frame". A window's frame is simply represented by the title of the window and all of the operating-system dependent window commands are ignored.

- The books referred to in this text (and their abbreviations) are as follows.
  (ES7) *Elementary Statistics, 7th Edition* by Robert Johnson, 1996. Wadsworth Publishing Company. ISBN 0-534-24324-X.
  Note: all references to pages numbers in ES7 appear as "pg. xxx" while references to pages within this companion appear as "page xxx".
  (JSS) *JMP Start Statistics*, by John Sall and Ann Lehman, 1996. Duxbury Press at Wadsworth Publishing Company. ISBN 0-534-26565-0 (book), ISBN 0-534-26564-2 (JMP IN 3 Windows version), ISBN 0-534-26562-6 (JMP IN 3 Macintosh version).
  (JMP/IG) *JMP Introductory Guide, Version 3.1*, by SAS Institute, 1995. ISBN 1-55544-680-9.
  (JMP/UG) *JMP User's Guide, Version 3.1*, by SAS Institute, 1995. ISBN 1-55544-678-7.
  (JMP/SG) *JMP Statistics and Graphics Guide, Version 3.1*, by SAS Institute, 1995. ISBN 1-55544-679-5.

Chapter 1
Getting Started with JMP

If you read nothing else ...

This chapter gives you the basic information you need to get started with JMP IN. Either the discussion in JSS or the JMP/IG and JMP/UG documentation are more complete than our coverage in this chapter. This chapter's abbreviated coverage is meant to fill the gap if neither of these other sources are available. If you are already familiar with JMP, you can start with Chapter 2.

Menu overview and JMP help

To start a JMP IN session, double-click the JMP IN application icon or select its icon and choose **File>Open** from the system's main menu.

The JMP IN menu bar appears along with an empty untitled data table.

 File Edit Tables Rows Cols Analyze Graph Tools Window

File Edit Tables Rows Cols Analyze Graph Tools Window Help

As you can see the two versions, Macintosh (above) and Windows (below) are almost identical. The main difference is that the help system is accessed from the Apple () menu on the Macintosh and from the **Help** menu under Windows.

Apple and Help menus

?! **How do I ...? Whatzit?** Consult the software documentation for details on what each of these menus do. What, no documentation? No problem. The JMP help system is extensive and shows up in nearly every window and display. Therefore, there are many ways of getting help in JMP. Get accustomed to one of these ways and use it to obtain help. For help on anything, do the following:

- Under Windows, choose **Help>Contents**.
- On the Macintosh, choose **Apple>About JMP**.
- For information on what any menu does, click the **Main Menus** button in the "JMP Statistics Made Visual" window.

Other buttons include **Stat Guide**, **Data Table**, **Calculator**, **Analysis**, **Graph**, **Credits** and **Version**. Most any question you may have concerning how to use JMP IN will probably be answered though one of these buttons.

The **Stat Guide** button is of special interest. When you click on the Stat Guide button the JMP Statistics Guide window appears. This guide is an alphabetical listing of common statistics terms and tells you how to produce these statistics in JMP IN. See JSS pages 13-16 for more information on getting help within the JMP help system.

Other Help There are other ways to get help while inside a window.

? By choosing the question tool from the **Tool** menu, the question mark appears in the analysis display window. Clicking on any object (in a graph) or on any text (in a report) will bring help associated with the object or text. This help is called context sensitive help.

Help is available for any analysis display by clicking on the star popup on the bottom left corner of the window. This help relates to the contents of the window from which the help is requested.

Help When requesting an analysis a dialog window may appear. There is a help button in the dialog window to provide help on how to complete the dialog.

TOC & Index All of your books have a table of contents (TOC) and index. The index in the back is especially useful for finding topics of interest. If you can't find a topic, try thinking of a synonym.

File and Edit menus

Briefly, the **File** and **Edit** menu entries should be familiar to you. However, the instructions on how to use the **File** menu for **Reading data from the student disk** are on page 4. You can also want to explore **File>Preferences** to tune JMP IN's defaults.

Tables, Rows and Cols menus

The next three menus work on JMP data tables. The **Tables** menu modifies or creates a new JMP table from one or more already open tables. The **Rows** menu modifies the rows in the data table and the **Cols** menu modifies the data table columns.

Remember the only "rule" of a graphical user interface:

To tell your computer to do something involves two steps:

1) Select an object, and
2) Choose an action that you want to perform.

If you can't do an action, it is probably because you have not **first** selected the object to act upon.

That is, if you want to modify a row and you see that what you want to do is grayed out in the **Rows** menu, select the rows you want to modify first. If you want to modify a particular column, you must highlight the column header before you can choose a command in the **Cols** menu. If you want to change a data table, JMP IN may need the data table to be the top-most (active)window before the **Tables** menu commands can work.

Analyze and Graph menus

The **Analysis** and **Graph** menus are the launch pad for JMP analysis or graph platforms. Both menus use the Y and/or X data in the top-most (active) data table to create a window that shows various graphical and tabular displays. See the JMP help system or your documentation for details.

Tools menu

The default tool is the arrow tool. If you want context-sensitive help on the contents of any window created by **Analysis** or **Graph** menu commands, choose **Tools>?** and click on the window object you don't understand. There are other tools useful in a variety of situations. Check the documentation or help system for more information.

Window menu

Since each analysis platform places another window on your screen, after a analyzing data for a while you will have a number of windows. To help you manage this bounty of windows, check out the **Window** menu. It's a good idea to close windows (**File>Close**) you are not using; just remember that if you've closed a window and need it again, you'll have to go through the menu commands and options to regenerate it.

JMP Data Tables

You need to know how to read data into JMP IN, how to save data to disk, and how to record your resulting analyses. You may also need to know how to rearrange a data file if it is not organized properly.

Reading data from the student disk

To begin, JMP IN needs data. The most common method for obtaining data is to read it from disk into JMP IN's working memory (into RAM). There are three kinds of files that JMP IN can read: SAS transport files, JMP data files, and unformatted text files. Data on the disk *Data disk for Elementary Statistics, Seventh Edition* in the back of ES7 is in several formats but *all you need is the data in SAS transport format*. Follow the instructions on the disk to uncompress the files and copy them to your hard drive. All of the SAS transport files have the extension **.xpt**, even on the Macintosh.

These student files are not saved in JMP format so you do <u>not</u> use the Open command in the File menu to read them. To read a SAS transport file into JMP IN's memory, you must *import* the file.

- Choose **File>Import>SAS transport file**, and select the file ?????.Xpt in the dialog. You may have to navigate through the directories to find it. Click the **OK** button (Windows) or **Open** button (Macintosh) on the open-file dialog.
- Check to see that the number of rows in the table is equal to the number of observations described in the textbook.

?! Some of the SAS transport files on disk have a last row that has strange values. To remove the last row there are two steps:

- highlight the row by clicking the space that contains the row number. The row number block turns black, indicating that you have selected it.
- Choose **Rows>Delete Rows**.

?! You may also notice that some data values in the SAS transport files on the student disk appear strange. That is, 16.0 may show up as 15.99999 in the JMP data table. You can safely ignore these appearance differences. Or, if it makes you feel better, you can type in the rounded values (see **Modifying a JMP data table**, below, for how to do this).

Reading JMP data tables

A JMP data file is a specially formatted file saved on a floppy or hard disk. Under windows these files usually have the extension **.jmp**. On the Macintosh an extension is not required. There are two ways to read a JMP data file into JMP IN's memory. One way is to use the JMP IN **File** menu.

- Choose **File>Open**, and select the file ?????.JMP in the dialog. You may have to navigate through the directories to find it. Click the **OK** button (Windows) or **Open** button (Macintosh).

? Can't see your JMP file under Windows? The file may have been named improperly. Try changing **List Files of Type:** from **JMP Files (*.JMP)** to **All files (*.*)**. The **show all files** check box can also help if you have a similar problem on the Mac.

The other way to read a JMP file is to use the fact that your operating system (either Win or Mac) knows who created a file. JMP files were created by JMP (or JMP IN). If you open a JMP file while on the Mac desktop or while under Windows, the system knows to open it using JMP IN.

- Double-click any JMP file document icon. 

Creating a JMP data table from scratch

The small data files in ES7 are not on the student disk so you need to type the data into JMP IN directly. Since the JMP data tables have a spreadsheet layout, it is easy to build a data table from scratch. JMP data tables have rows and columns, also called observations and variables. Consider carefully how you organize a data table. Generally, it is a good idea to lay it out so that independent observations are in separate rows and their characteristics are recorded in different columns.

To build a data table there are just Four steps: Create a new table, define the columns, add rows, and enter the data:

- To create a new table, choose **File>New**. An untitled window appears with one column and no rows.
- To define the columns you have to do two things: 1) Make the columns appear and 2) set their characteristics appropriately. The easiest way to create many columns at once is to choose **Cols>Add Columns**. Or, if you want to make one column at a time, choose **Cols>New Column**. If you do the

latter, you can set a column's characteristics through the New Column dialog. Then set each column's characteristics by double-clicking with the open cross cursor above the column's name. The *Column Information* dialog appears.

There are four things you must pay attention to in these dialogs:

- **Col name:** is the column's name. Name columns something that make sense to you. Use up to 31 characters, including spaces and special characters like $#%.

- **Data Type:** is a popup menu with two relevant choices, **Numeric** and **Character**. There is a third choice, **Row State**, that we'll not cover.

Which Data Type?

If you want to do any arithmetic on a column it must be numeric. That is, if you have some financial figures (like $12,243.15) and want to calculate totals, you must define the column as numeric and enter the data as: 12243.15 (no dollar signs or commas; the decimal is OK). On the other hand, if you have subject numbers (like 408-91-2093) it is unlikely that you need to do any arithmetic on this column. So, you could define the column as character (and enter the data as: "408-91-2093", without the quote marks) or numeric (and enter the data as: 408912093, without the dashes). Columns that have character values (M and F, for instance) must be defined as **Data Type: Character**. The only thing you should type in a **Data Type: Numeric** field are numeric digits, a decimal, and a leading negative sign. Notice that you can tell a column's data type by looking at the spreadsheet. Numeric columns are right justified while character columns are left justified.

? What about dates? Dates should probably be defined as **Data Type: Numeric** with **Format: Date and Time**. Then dates are displayed as *mm/dd/yy*.

- **Field Width:** is the number of characters that fit in a column. It is only relevant for character columns. Change the default from 8 to whatever is necessary (up to 255). Leave this characteristic alone for numeric columns.

- **Modeling Type:** describes how you want JMP IN to treat this column in analyses. See the section **Absolutely Core JMP Concepts: Modeling Type** on page 12 in this chapter. A *continuous* modeling type implies that you want JMP IN to use the numeric values directly. The categorical modeling types are ordinal and nominal. Both imply that there are discrete, categorical values that name the levels or different groups. *Nominal* implies that these groups are just "different" while *ordinal* implies that their order is important. The order of a character column is alphabetical and the order of a numeric column is from smallest to largest. You can always change modeling type later so don't feel like this decision is cast in stone.

- To add as many rows as you need, choose **Rows>Add Rows**. A dialog appears. Type a number indicating as many rows as you like. You can always come back to the **Rows** menu and add more rows later.
- To enter the data into the cells is easy. Move the cursor to a cell by clicking on the cell. The I-beam cursor becomes a blinking vertical bar, which indicates you can start typing.

Typing in Data

The Tab key moves the cursor one cell to the right. Shift-Tab moves left.

The Return key moves the cursor down one cell. Shift-Return moves up.

If you make a mistake, just drag the I-beam cursor over the incorrect entry and type over it.

Don't type quote marks in character columns or special characters in numeric columns. That is, if a column's name is Gender, type M and F, not "M" and "F". If you want to type numeric values into Money, type 1040.13 not $1,040.13.

If you don't know the value for a cell, leave the cell alone. Missing numeric values display as • on the Mac and ? under Windows. A missing character value displays as nothing. The cell looks empty. Do not type a space for a character missing value. A space (the character generated when you strike the big wide bar at the bottom of the keyboard) is a non-missing value.

If you have dates and you have defined the characteristics as described above, you can type June 13th, 1995 as any of the following: 6/13/95, 6-13-95, 6 13 95, 06/13/95, 06-13-95, 06 13 95, 6/13/1995, 6-13-1995,

6 13 1995, and probably some others. You get the idea. They all display as: 6/13/95. Be careful about reversing months and days; Is 2/3/95 February 3rd or March 2nd? The default ordering, *mm/dd/yy*, is set by your system. Virtually everyone else in the world except Americans think this is a nonsensical ordering. Either *dd/mm/yy* or *yy/mm/dd* makes much more sense. JMP IN will handle whichever way your system is configured.

If you are typing in a lot of data, save the file to disk often. You never know when there may be some power outage.

Using Copy and Paste

If you already have your data keyed into a spreadsheet program like Excel you do not need to type it into JMP IN again. Roughly, all you need to do is copy the data from the spreadsheet program, create a new data table in JMP IN, and paste the values in.

There is a trick that lets you automatically create column names in the new JMP data table; it has to do with whether the first row of information in the spreadsheet is the column names. Here are the steps in detail:

- Start in the program that has your data, in this case Microsoft Excel. **Open** the file containing the data. The data is shown with a split screen so that you can see that there are 51 rows of information, including the row of column names.

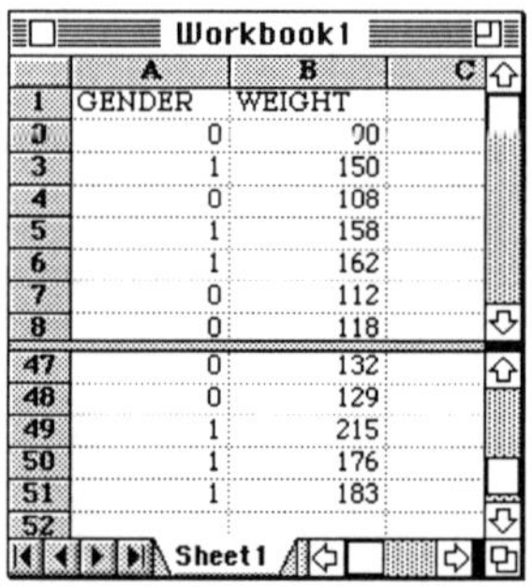

- Select the entire set of data by Shift–clicking.

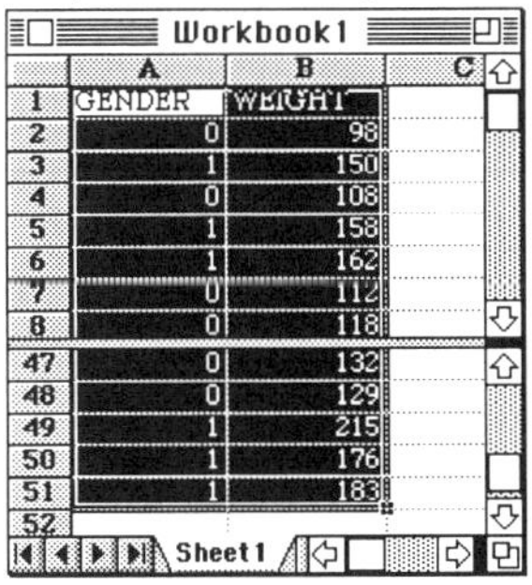

- Ask yourself: What is in the first row of the spreadsheet? The two usual answers are a) column names or b) data. Remember your answer.
- Choose **Edit>Copy**. The data and column names are on the clipboard.
- Now go to JMP IN. You may need to Quit or Exit Excel first and then launch JMP IN.
- Choose **File>New** to create an empty data table.
- Recall your answer to the question above.

? If your answer was a) column names were in the first row of the spreadsheet, Hold down the Option key (on the Mac) or the Alt key (under Win) *and* Choose **File>Paste**.

? If your answer was b) just data were in the first row of the spreadsheet, Choose **File>Paste**.

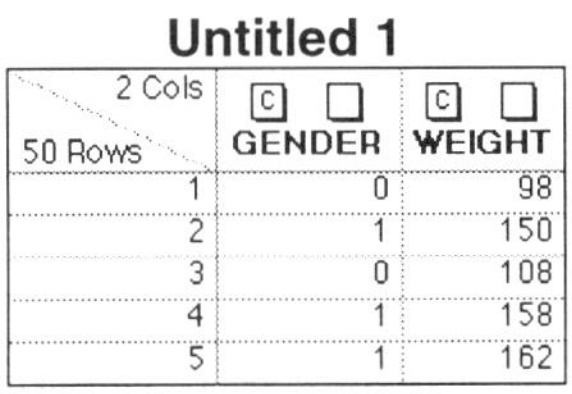

If column names were saved in the Copy step, they are used as column names for the JMP IN data table. When you held down the Option/Alt key, you were warning JMP IN not to expect data in the first row of information. Normally, JMP IN would assume that what is in the clipboard is just data. If you only copied data, JMP IN names the columns Column 1, Column 2, ... See JSS pages 13-16 for more information on getting help within the JMP help system.

Check that the JMP data table looks right. In particular, check to see that the columns that had numeric values are right justified and the columns that had character values are left justified.

You may not be done yet. If any of the columns identify groups you must change the modeling type. See the section **Absolutely Core JMP Concepts: Modeling Type** on page 12. In addition, you may need to slice and dice the data table to arrange it so that it is suitable for analysis using JMP IN. As Sall and Lehman say on page 33 of JSS "Never underestimate the diabolical convolution of data that can appear in an electronic table. ... Always check your data with common sense." This fact can not be over emphasized. Just because someone gave you data to work with (even us) does not mean it is right. Murphy's law of data: there is always one more mistake in the data.

Modifying a JMP data table

Everything you probably need to manipulate your data table to take it from however it is arranged to however it needs to be organized is in the **Tables** menu. These commands are briefly summarized below. If you have trouble accomplishing your tasks with any of these commands, you should read the software documentation. See Chapter 2 in JSS on juggling data tables. Also, do not forget that each of these commands has a help button on its dialog.

Tables> Subset Use the **Subset** command to extract rows and columns from one table and create a new table with just the selected subset. If no rows are selected the subset contains all rows whereas if you select a subset of rows only those rows are extracted. If no columns are selected the subset contains all the columns, whereas if you select a subset of the columns only those columns are extracted.

Tables> Sort Use the **Sort** command to rearrange the rows so that they are sorted in order by the values in one or more columns. See page 32 for an example.

Tables> Stack Columns Use the **Stack Columns** to take data that are in two or more columns and stack them on top of one another. The stacked table has two new columns: the name of the columns that were stacked and the data of the columns that were stacked. See the exercise beginning on page 99 for an example.

Tables> Split Columns Use the **Split Columns** command to do the reverse of **Stack Columns**. A common reason to split columns is when values are stored in separate rows but they are actually paired responses on the same subject.

Tables> Transpose Usc thc **Transpose** command to transposc rows into columns and columns into rows.

Tables> Concatenate Use the **Concatenate** command if you have two data tables with the same columns and you want to combine tables by adding the rows of one table onto the end of the rows of another.

Tables> Join Use the **Join** command if you have two data with comparable rows but different columns and you want to combine tables by adding the columns of one table onto the columns of another table. Also use **Join** to combine two tables that have a matching column that should be used to hook the observations together.

Saving a JMP data table

If you have imported a SAS transport file, modified an existing JMP data table, or created a new data table from scratch you probably want to save it on disk so you can read it later.

- Choose **File>Save**, and complete the dialog. JMP file names on the Macintosh and Windows 95 can have any length. JMP files under Windows should have the extension .jmp. To keep things straight, it is probably a good idea use some sort of consistent naming strategy. That is, if you import the file TAB02-02.Xpt, you should probably save it as TAB02-02.JMP. If at any point, you want to change the name of the file or if you want to save it in a different directory or disk, choose **File>Save As**.

Absolutely Core JMP Concepts

You have probably noticed by now the two small boxes over the column names. The box to the left identifies the modeling type and the box to the right contains the column role. As you begin to understand how JMP "thinks", you begin to understand the importance of the modeling type and the column role. For example, assuming the model type is specified correctly, JMP only does the most appropriate analysis when an analysis is requested.

In this section we define the different modeling types and column roles. Also, the various types of reports are introduced by briefly introducing the **Analyze>Distribution of Y** platform.

Modeling Type

As Section 1.2 in ES7 points out, data basically come in two "flavors." There are lots of equivalent terms for these flavors: qualitative and quantitative, attributes and numerical, discrete and interval. But the JMP terminology is: categorical and continuous. JMP IN calls this distinction modeling type. This is because JMP IN needs to know how to treat the data when it does an analysis.

Why do you care? From a strictly practical point of view (goal: you want to get your homework done), if you use the data on the student disk as-is, you will not succeed (result: your homework will be wrong). Here is the situation you must deal with: all of the data on the student disk are numeric but not all of the data are continuous..

Continuous Modeling Type JMP begins by looking at data and presumes that all numbers are *continuous*. If a column is continuous. then possible values include any number you can think of (and then some). Another way to think about continuous. data is that it can arise from an interval. For example, weight can be thought of as coming from the numeric interval from 0 to 300 (or so). If a column is , it makes sense to do things like take an average. Weight is continuous. Consider the following portion of a data table.

TAB02-02

2 Cols / 50 Rows	C GENDER	C WEIGHT
1	0	98
2	1	150
3	0	108
4	1	158
5	1	162
6	0	112
7	0	118
8	1	167

JMP IN assumes all numeric values are continuous. If you ask JMP IN to calculate the averages for these two columns—and don't worry about *how* to do that yet—JMP IN obliges and produces the following reports.

GENDER

Moments	
Mean	0.60000
Std Dev	0.49487
Std Error Mean	0.06999
Upper 95% Mean	0.74064
Lower 95% Mean	0.45936
N	50.00000
Sum Weights	50.00000

WEIGHT

Moments	
Mean	150.6400
Std Dev	30.5258
Std Error Mean	4.3170
Upper 95% Mean	159.3153
Lower 95% Mean	141.9647
N	50.0000
Sum Weights	50.0000

One of these tables makes sense (the mean weight is about 150 pounds) and one does not (the mean gender is 0.60). If you don't pay attention you can get nonsense answers.

Nominal Modeling Type The second "flavor" of data is *categorical*. Data are categorical when the data can take only a discrete (limited) number of values. Gender and race are examples of *nominal* categorical data. Notice that gender can only be expressed as two values, male or female. On the student data disk these values are usually coded as 0 and 1, or as 1 and 2. In the same way, race can be defined by three or four nominal groups such as Caucasian, African-American, Hispanic and other. JMP IN needs a hint from you though. In the race example, if race is coded 1, 2, 3, 4, is it continuous? One way of deciding if data are continuous or categorical is by considering if an average is appropriate for the data. Because the idea of calculating an average gender or race make absolutely no sense, both gender and race should be marked for JMP IN as the nominal modeling type. If data have character values JMP IN automatically presumes that the column is nominal.

Ordinal Modeling Type In order to most appropriately work with categorical data, JMP makes one further distinction. JMP IN uses two categorical modeling types: ordinal and nominal. *Ordinal* data are discrete, qualitative, categorical data in which there is an ordering to the values. Examples of ordinal. data include grade level (freshman, sophomore, junior, and senior) and the number of speeding tickets (1, 2, 3, and 4 or more). On the other hand, nominal data have no natural ordering; gender and race are examples of nominal data. Remember that in nominal data the values are just "different." In data the categories are ordered. If you are not sure, try the nominal. modeling type; it makes the least assumptions. You can always change your mind later.

How do you select the modeling type? The little boxes above the column names in the data table are called popup menus. When you put the tip of the arrow cursor in the left box and hold down on the mouse button, the options of [C] **Continuous**, [O] **Ordinal**, or [N] **Nominal** popup. Drag the arrow to the desired modeling type and release mouse button; that is how you select modeling type. You can change the modeling type any time the data window is the top-most (active) window.

TAB02-02

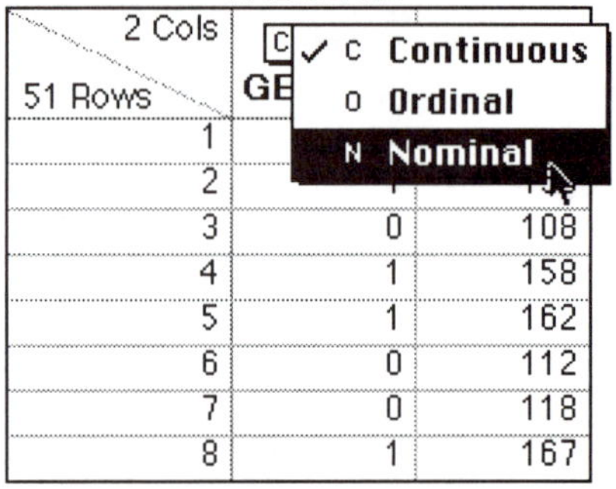

Column Roles

Like the modeling type popup menu, the column roles popup menu provides JMP IN with information on how to treat the column in an analysis. A brief description of each modeling type follows:

None A column role of **None** indicates that the column is to play no role in the upcoming analysis.

[X] **X** indicates the column is a predictor variable. Other names for variables that have this role in an analysis are independent variable., classification variable or explanatory variable. In bivariate displays the X values appear on the horizontal axis of the plot.

[Y] **Y** indicates the column is a response variable. Sometimes this is called the dependent variable In bivariate displays the Y values appear on the vertical axis of the plot.

[W] **Weight** The column with the **Weight** role supplies weights for the analysis. Weights are usually a number between 0 and 1 that corresponds to the importance of data. This role is not used in any example in ES7.

[F] **Freq** A column with the role of **Freq** tells JMP IN the number of occurrences of that row of data. Normally, each row of the data tables "counts" as n=1. If you have a column with frequency counts, when you use the Freq role the column counts as n = the value of the **Freq** variable.

Label indicates which column should be used to label the observations. JMP IN's interactive graphs and plots allow you to point and click on a data point . Without a Label, column the row number appears in the plot. If you mark a column as the Label column, the information in the label column appears when you click on a point.

It is extremely important to be familiar with these column roles. Assuming that the column roles are specified correctly, JMP IN can begin producing the correct analyses by simply choosing an entry in the **Analyze** or **Graph** menu.

How do you select the column role? You select column roles in one of two places: In the data table or in an analysis dialog. In the data table the popup menus above the column names on the left show a column's role. When you put the tip of the arrow cursor in the right box and hold down on the mouse button, the options popup. Drag the arrow to the desired column role. You can change the column role any time the data window is the top-most (active) window. Specifying column role in this manner affects the *next* analysis.

TAB02-02

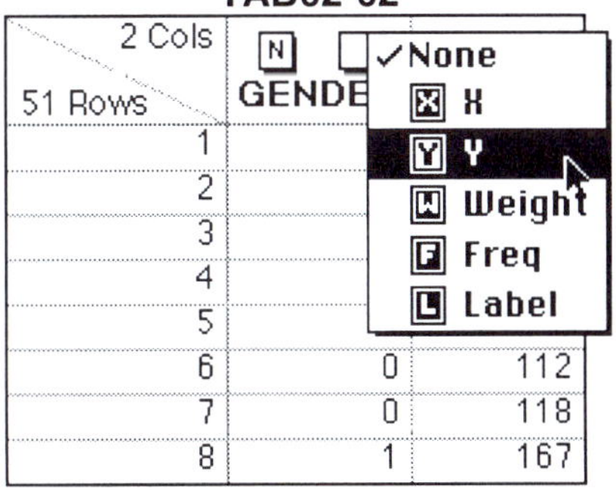

The other way to select the role for a column is to choose an analysis first. With no roles specified in the data table, if you tell JMP IN that you want to do a particular analysis a dialog appears asking you to specify which columns serve in which role. Select the columns from the scrolling list of column names and press the button corresponding to the role you want the column to take.

Analysis Platforms

There are a variety of analyses that JMP IN can produce through the **Analyze** and **Graph** menus. The results of these menu selections are called *JMP platforms* and each platform provides a different analysis or display. In order to demonstrate these menus, the simplest of the **Analyze** menu sections, **Distribution of Y** is shown next. To request the **Distribution of Y** platform, click on the **Analyze** menu and select **Distribution of Y**.

Depending on the modeling type of the Y columns of the active data window, JMP produces the appropriate graphical displays and text reports. For categorical. (ordinal. or nominal.) columns, a histogram, mosaic chart (stacked bar chart) and a frequency text report are produced. For continuous. columns, a histogram, outlier box plot, quintiles report and moments (means and variances) report. The details of the **Distribution of Y** platform are contained in the section **Describing Single Variables** in **Chapter 2 Descriptive Analyses and Presentation of Single-Variable Data**.

An Overview of the Analysis Platforms In this section, each of the platforms from the **Analyze** and **Graph** menus are introduced. Because of the graphical emphasis of JMP IN, many interactive graphical displays are produced when you launch an analysis platform. The platforms in the **Analyze** menu also produce text reports. The platforms in the **Graph** menu primarily produce graphical displays. Not all of the platforms are used in ES7.

Analyze Menu There are six platforms under the JMP IN **Analyze** menu in and they are approximately ordered from simplest (top) to more complicated (bottom). There are seven platforms In JMP. The seventh platform In JMP does cluster analysis and is not discussed in this companion. A brief description follows for each platform.

Distribution of Y Univariate analyses are produced which includes histograms, outlier box plots, and descriptive statistics like means, variances and quantiles for continuous columns. For ordinal or nominal columns histograms, mosaic charts (stacked bar chart) and frequencies are shown.

Fit Y by X Bivariate (two columns at a time) analyses are produced for each pair of the X and Y columns. Depending on the modeling type of X and Y, this platform does regression, analysis of variance (or t-tests), contingency table analysis or logistic regression.

Fit Model This platform is a general model fitting platform which allows multi-variable (more than one X column) analysis or multivariate (more than on Y column) analysis or both. This platform is not used in ES7.

Fit Nonlinear This platform fits nonlinear models. The **Fit Model** platform only models "linear" models (linear combinations of the X's) but the **Fit Nonlinear** models more general forms of the X's. This platform is not used in ES7.

Correlation of Y's This platform describes the relationship between many Y columns by producing correlation coefficients and scatter plots.

Survival This platform models survival (or time to event) data. It is called survival because the "event" is usually death, so it is time to death analysis or survival analysis. This platform is not used in ES7.

The platforms, as described in this introduction, may appear to be fixed and ridged. Nothing could be further from the truth. The platforms are just starting points to begin an exploration. Each platform is designed for adaptation and exploration. The default display can be modified to suit a presentation or to find features of the data.

More detail for the **Distribution of Y**, **Fit Y by X** and **Correlation of Y's** platforms follow in the coming chapters. The **Fit Model**, **Fit Nonlinear** and **Survival** platforms are not used in this companion. For more information, see JSS or JMP/SG.

Graph Menu The menu contains some specialized plots frequently used by statisticians. (A statistician is sometimes described as someone that does not have the personality to be an accountant.) Each platform is described below.

Bar/Pie Charts This platform produces various charts such as bar, pie, line and needle.

Overlay Plots This platform produces overlay plots. These plots get their name from "overlaying" 2 or more Y columns (along the vertical axis), across the one X column (along the horizontal axis).

Spinning Plot This platform produces a three dimensional scatter plot that can be rotated to see depth.

Pareto Charts A pareto chart displays bars that show the frequency count in descending order with a line showing the cumulative counts.

Control Charts A control chart is a specialty chart used to monitor a process (such as a part coming off an assembly line) and determine whether the process is "out of control."

More detail for the **Bar/Pie Charts**, **Overlay Plots** and **Pareto Charts** platforms follow in the coming chapters. The **Spinning Plot** and **Control Charts** platforms not used in this companion. For more information, see JSS or JMP/SG.

Interacting with the Analysis Displays

The JMP IN interactive analysis displays are very much alive. These displays were designed to help you discover important features of your data and lead you to a correct analysis. In what follows, a description is given of some important buttons, tools and menus.

Graphs and Reports

Every graph produced in JMP IN has interactive features. For example, by clicking on (and hightlighting) one of the bars in a histogram, parts of the bars in other histograms are highlighted. In other words, the proportion of the selected bar in the first histogram is highlighted in the bars of the other histograms. Scatter plots work in much the same way; by clicking on a point or on several points in a scatter plot, points in other scatter plots or histograms are highlighted.

At the same time points in a scatter plot or bars in a histogram are highlighted, the corresponding entries in the data table are highlighted. You may wish to subset, delete, hide, etc. these rows. Or you may be want to know the row number of an odd looking point in a scatter plot. JMP IN is designed so that the various displays and data tables are dynamically linked.

Using the JMP Tools

JMP provides the user with various tools that have functionality in the analysis displays. The tools are selected by clicking on the **Tools** menu and selecting the tool. The arrow cursor is changed into the selected tool when placed in an analysis display. Each tool is described below.

The arrow tool is the standard tool used for most actions. In a report window, use it to select points and histogram bars, select popup options, and reveal text reports. In the data table, use it to select rows or columns, select text for editing, or make assignments in the role or modeling type popup menus.

The following tools work only in the report windows, not the data table.

The hand tool allows direct manipulation or *grabbing* in graphical displays. This tool is demonstrated in Chapter 2.

? The question mark tool allows access to JMP IN's context sensitive help. Context sensitive help is discussed in the next section.

The brush tool is especially helpful when you want to select a group of points in a plot. By selecting the brush tool and clicking in a plot, a rectangle appears and can be dragged over points to select them. The rectangle can be enlarged by using the Alt-key modifier in Windows and the Option-key modifier on the Mac.

The crosshair tool, most useful in scatter plots, is a movable set of axes for which the middle of the crosshair is the origin. When you click in a plot, the horizontal and vertical coordinates from the original scaling are shown as long as you hold the mouse button down.

The scissors tool is used to select areas of a graphical display to be copied and pasted to a report window (usually in a word processor document). Selection is done by clicking on one corner of the display and dragging diagonally until the area of interest is selected.

The lasso tool allows the selection of irregular shapes of points in plots by dragging the "lasso" around the desired points. The lasso closes when the mouse button is released, thereby selecting the points.

The magnifying tool zooms into the area of a plot by clicking. With each click, the image is enlarged by about 25%. By holding down the Alt-key in Windows or the Option-key in the Mac and clicking on the plot again, the plot is restored to its original scaling.

The annotate tool places a "sticky-note" in the report window. Type text into the note and use the optional lines extending out of the edge of the note to point to points or enhance a graphical display. To remove a note, drag it out of the window.

Border Popup Menus

Along the bottom, left corner of an analysis display window in JMP, there are three popup menus: check, dollar and star. Regardless of the analysis display, the popups have the same function. Each are briefly described.

The check popup menu is best described as a whole window modifier. The entries of the menu vary depending on the analysis display. When an option from this menu are selected the option applies to all displays in the window.

The dollar popup menu saves results to the data table.

The star popup menu provides options that are common to all platforms. For example, help is available through this menu for the platform that generated the display. A more complete description of help is discussed in the next section.

Interior Popup Menus

When you see the interior popup menu in an analysis window, there are options in the menu for that display. These options apply to that display only, in contrast to the check popup menu for all displays in the window. The interior popup menu icons are located to the right of a report name or along the bottom of a display.

Report Reveal Buttons

Text reports are also interactive in that they can be hidden or revealed. By clicking on the report name **t-Test** the report is alternatively hidden or revealed. These report titles are called *reveal buttons*.

So, there are six facets of JMP IN covered in this chapter: the menus, the data table, core concepts of modeling type and column roles, launching analysis platforms, and interacting with the displays. If you've gotten this far, you have some results you want to put down on paper.

Recording Your Results

One way to record your findings is to write them down on a piece of paper. Pencils still work. However, it is very simple to get much more impressive displays out of JMP IN.

Printing

You can print the top-most (active) window. Make the display look like you want—select things you want to highlight, resize the figures as you desire, and reveal any text reports of interest—and choose **File>Print**.

Journaling

To get a disk version of your results, you do what is called JMP *journaling*. The process is similar to printing. Make the contents of a window look like you want and choose **Edit>Journal**. All the tables and figures are placed in a journal window. Do not close it yet. If you interested in saving the results of any other windows on disk, make them the top-most window, make them look like you want, and journal again. Every time you do this more tables and figures are appended to the bottom of the journal. You can also type into the bottom of the journal window. When everything you want is in the journal window, save and close the window.

Other sources of information

We know it is fashionable not to read manuals or books, but perhaps if you have gotten this far you might profit from one piece of advice: Don't read manuals or books until you need to. Pick them up when you need to and find the pieces you are ready to digest at the moment. You may even stumble across something that helps you later. If you are having problems it is a good bet that somewhere in this book, ES7, JSS or the JMP documentation is your answer.

If you want to "learn by doing" the JMP/IG can't be beat. If you want another point of view of the ins and outs of JMP, read JMPing in with both feet in JSS.

The only area we know of where you almost certainly need to read some documentation is if you do anything more than simple work with the JMP IN calculator. Calculator adventures in chapter 3 of JSS is good, as is the definitive reference in chapters 5 and 6 of the JMP/UG.

Chapter 2
Descriptive Analyses and Presentation of Single-Variable Data

Describing Single Variables

When beginning an analysis of data, the first step is to understand the values of each variable. This is done with the first analysis platform, **Distribution of Y**. Recall that with JMP IN, a Y column is the "column of interest" or the response variable. In this chapter the Y role is the former. To "analyze the distribution of Y" means that you are interested in the values stored in a column.

Note that the use of "Y" to mean "an arbitrary variable name" is simply JMP IN's convention. In ES7, X is used. In this chapter, when you are only interested in a single variable, it does not matter which symbol is used. However, when thinking ahead to more complicated analyses, it is convenient to make a distinction between the Y and X role. As you will see in the next chapter when you are interested in the relationship between two variables, this is an important distinction. So, whereas ES7 uses X because of algebraic tradition, JMP IN uses Y to insure data analysis consistency.

John Sall and Ann Lehman discuss the topics in this chapter in JSS Chapter 5, *Univariate Distributions: One Variable, One Sample*. In addition, Pareto charts are covered in JSS Chapter 15, *Statistical Quality Control*.

Using Distribution of Y to describe a variable

The **Distribution of Y** command in the **Analyze** menu includes the ability to generate a wide variety of graphical and tabular summaries of columns. JMP IN's default display for continuous columns include a: histogram, box plot, and numeric summary of descriptive statistics. Options for continuous Y's include: stem-and-leaf plots, superimposing a normal or smooth curve on the histogram, testing for normality, and testing for a specific mean. Nominal or ordinal variable displays from **Distribution of Y** include a: histogram, mosaic chart, and table of frequency counts. Analyzing qualitative variables is covered more fully in ES7 Chapter 11. For all displays you can modify the cut-points in a histogram, change axis scaling and labeling, and save rescaled values.

- To obtain these descriptive displays, make the data table of interest the top-most window in JMP IN and choose **Analyze>Distribution of Y**. If you have marked Y columns already, you see the default **Distribution of Y** display. If not, select the columns of interest in the dialog that appears, and click OK.

The first example uses TAB02-5, which has a single column, SCORE. Because the column of interest is numeric, any modeling type can be used. The default modeling type is continuous.

Graphical displays

The top of the resulting window contains a button with the column's name, an ▶ option popup next to the column name, and a graphical display that shows a grouped histogram and a box-and-whiskers display. Two text reports are also visible below the graphical display and are discussed later.

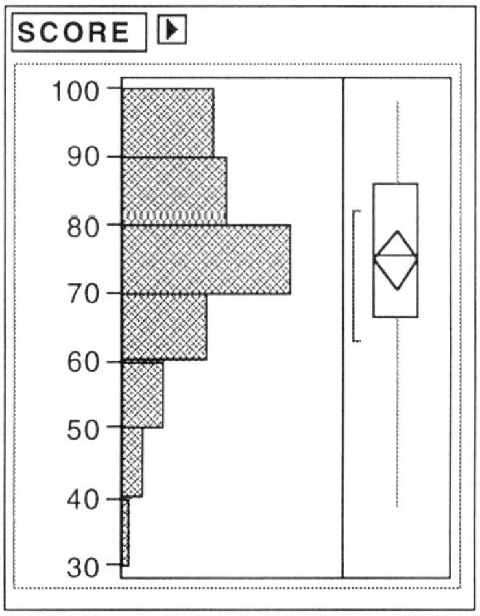

? If your display has a histogram but does not have the box-and-whiskers display, you specified that the modeling type of the variable as nominal. or ordinal.. If you meant to treat your variable as though it were continuous., close the **Distribution of Y** window, go back to the data table to change the column modeling type to continuous, and rerun the analysis.

Histogram scaling In the display for **TAB02-5** JMP IN has determined the vertical axis, which ranges between 30 and 100, with class boundaries every 10 units. When a column's modeling type is continuous, JMP IN assumes that you do not want one bar per value. The assumption is that you want to group ranges of values together and plot a *histogram* (or, what is sometimes called a grouped histogram). Values within these 10 unit boundaries are grouped together and the length of the bar is proportional to the number of observations in each bar. Equivalently, the bars are proportional to the percentage of observations in each bar. The tick marks on the axis appear at each boundary (as in ES7 Figure 2-11, pg. 59), not at the midpoint of the ranges (as in ES7 Figure 2-10, pg. 58).

How do you change the cut points? You can either do it interactively with the hand tool or manually by filling in a dialog.

To interactively change the width or number of bars in a histogram, choose **Tools>** and then click-and-drag inside the rectangle enclosing the histogram. Dragging to the right makes more bars and makes them narrower; dragging to the left makes less bars and makes them wider. Dragging up and down keeps the same width of the bars and only changes the cut points.

After dragging right all the way:

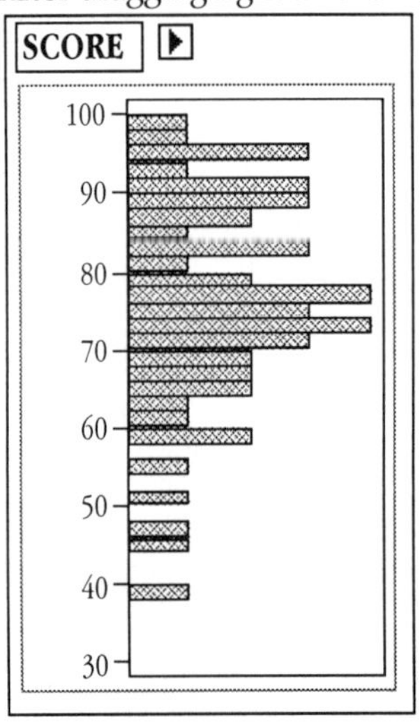

After dragging left all the way:

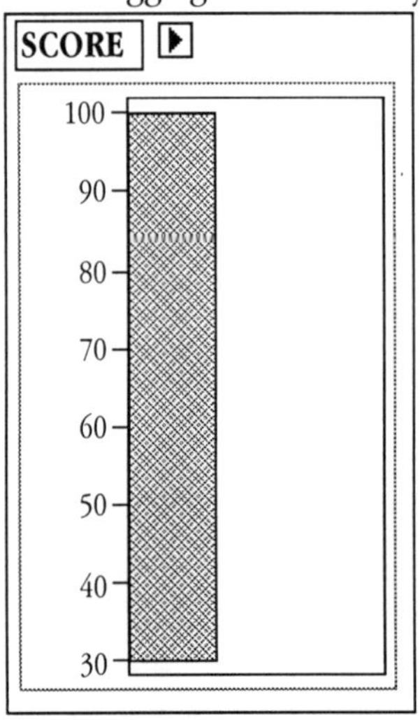

Another way to change the scaling of the axis or histogram cut points is to double-click on the numerical axis area. The *Axis Rescale dialog* appears. If you specify no minor tick marks, there is one histogram bar per increment. If minor tick marks are used, one bar per minor tick appears.

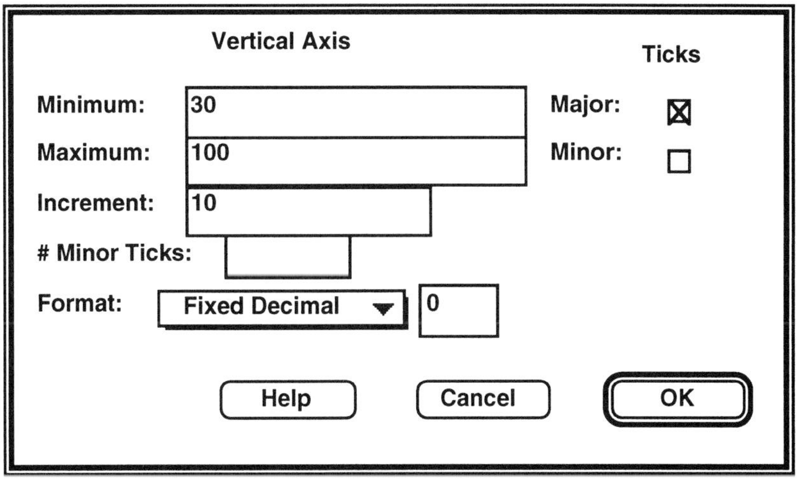

After specifying **Minimum: 35**, and **# Minor Ticks: 1** the histogram appears as below.

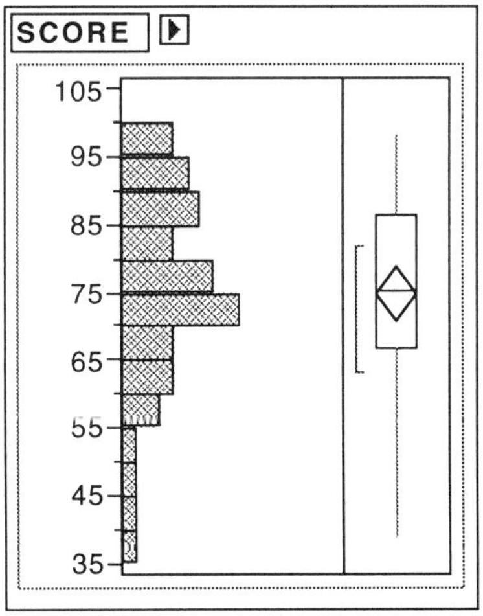

The box-and-whisker plot To the right of the histogram, an *outlier box plot* appears. This is similar to ES7's *box-and-whisker display* in Figure 2-30 on pg. 99. The box encloses the two *quartiles* and the line in the middle of the box represents the *median* value. The width of the box is called the *inter-quartile* range. The dotted whiskers extend to the smallest and largest value in ES7 Figure 2-30, but in JMP IN the whiskers do not extend to the extreme values. The JMP IN outlier box plot whiskers extend to 1.5 times the inter-quartile range. If there are any data values beyond the limits of the whiskers, these potential *outliers* appear as a dot. JMP IN also displays a diamond representing the 95% confidence interval about the mean.

Text reports for continuous columns

Below the histogram, two reports appear. The **Quantiles** report contains the numerical values plotted in the outlier box plot. The **Moments** report contains estimated mean, standard deviation, standard error of the mean, the 95% confidence interval for the mean, and the number of observations with non-missing values.

Quantiles		
maximum	100.0%	98.000
	99.5%	98.000
	97.5%	97.725
	90.0%	93.800
quartile	75.0%	86.500
median	50.0%	75.500
quartile	25.0%	66.750
	10.0%	55.300
	2.5%	40.375
	0.5%	39.000
minimum	0.0%	39.000

Moments ▶	
Mean	74.92000
Std Dev	14.19520
Std Error Mean	2.00750
Upper 95% Mean	78.95422
Lower 95% Mean	70.88578
N	50.00000
Sum Weights	50.00000

Optional displays and reports

Optional displays are chosen through either a window border or window interior popup menu command. The ☑ popup menu appears in the bottom left of the **Distribution of Y** window. The ▶ popup option appears inside the window to the right of the column name, above the histogram. Optional numerical summaries can be chosen in the **Moments** report.

Common ☑ popup menu options If you want the horizontal axis in the histogram labeled, choose **Count axis** to display the frequency count or choose **Prob axis** to display the relative proportion. The **Horizontal Layout** option makes the column's axis go from left to right, instead of up and down. To view the distribution as it would appear on normal probability paper, choose **Normal Quantile Plot**. Another comparison to the normal distribution can be seen when a normal curve is superimposed on the histogram by choosing **Normal Curve**.

Common ▶ popup options To compare the observed mean to a hypothesized mean using either a z-test, t-test, or nonparametric test, choose **Test mean=value**. To add a stem-and-leaf text report, choose **Stem and Leaf**.

Moments ▶ popup options If you are interested in the sum of all values, skewness, kurtosis, or coefficient of variation, choose **More Moments** from the popup to the right of the **Moments** report button.

Pareto charts

To produce a Pareto diagram you need a data table with a column indicating the different categories you want to describe. This column has the Y role. The column can be either numeric or character and can be any modeling type, although the nominal modeling type makes more sense.

You can either have one row per observation or you can have one row per category group. For instance, in Illustration 2-2, Figure 2-3 there are 4 crime categories and a total count of 4168+1189+806+583 = 6746 observations. You can either have a data table with 6746 rows (one per instance of the category) or a data table with 4 rows (one per category group). If you choose the latter, you'll need a column to indicate the number in each group. This column, indicating the number of observations in each category group, should have the Frequency role. For instance, the steps to build a pareto chart like Illustration 2-2 is shown below.

Using a frequency count variable

- Check the organization of the data table; notice the modeling types and column roles.

ILL2-02

2 Cols / 4 Rows	N Y Crime	C F Number
1	Race	4168
2	Relig	1189
3	SexOr	806
4	Ethn	583

Creating a pareto chart

- Choose the **Graph>Pareto Charts** menu command.

  If you have already specified roles—the Y role (Y) and the Frequency role (F)— in the data table, the chart appears.

- If you have not specified roles, a role assignment dialog appears.
  - Select Crime >Y>.
  - Select Number >Freq>.
- Click **OK**.

  The default Pareto display below corresponds to Illustration 2-2 in ES7.

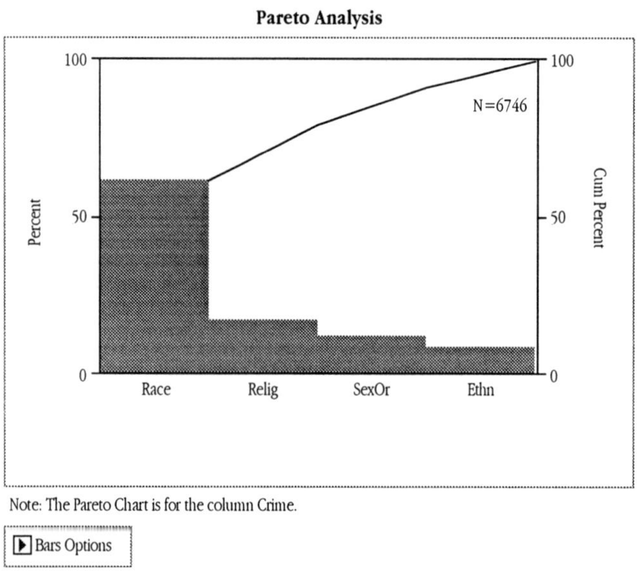

Modifying the display

The title "Pareto Analysis" can be edited, as can the footnote "Note: The Pareto Chart is for the column Crime." Notice that the JMP IN Pareto chart uses percentages on both vertical axes.

There are two sets of optional commands that can be used to change the default display. You can change the appearance of a bar by clicking on it and then choosing a ▶ **Bars Options** command. These commands allow you to combine, separate, and move the highlighted bars, and to change their color or pattern. You can also add a label corresponding to the bar value at the top of a highlighted bar. These commands act on the highlighted bar(s).

To tailor the appearance of whole Pareto display, use the ☑ border popup menu at the bottom left of the display window. Commands include options to change the display. The ☑ **% Scale** option changes the left hand vertical axis to display the counts of each category and makes the display correspond to Figure 2-3 in ES7.

Dot plots and ogives

Dot plots and ogives both plot observed values on one axis and a summary statistic on the other axis. Dot plots are simply a plot of an observed value versus the frequency count of observed values. Ogives are a plot of observed values versus the cumulative relative proportions.

Dot plots There are two ways to obtain dot plots in JMP IN. One way uses the Fit Y by X platform. A more precise method uses formulas in the JMP IN calculator. Let's do the easy way first.

Point plots, something akin to dot plots, are the default display in the Fit Y by X platform if the Y variables is continuous and the X variable is a categorical grouping variable. The easiest method for obtaining point plots is thus when you have multiple groups. The method for a single group involves a little trick.

Constructing multiple point plot displays

Figure 2-9 on page 47 of ES7 shows dot plots with a common scale. The intent is to compare the weights of two groups of students. The most straightforward way to view multiple dot plots is to use the Fit Y by X platform.

- Import the data table TAB02-02.XPT.
  - Choose **File>Import>SAS transport file**.
  - Select the file TAB02-02.XPT.

⁉ Something is amiss with this data table. Table 2-2 on page 45 indicates that there are 50 college students. Notice from the upper left triangle in the data table window, that there are 51 rows and 2 columns. It is always a good idea to look at the data before proceeding with analyses. Scroll down to the last row and you see the problem.

TAB02-02

2 Cols / 51 Rows	GENDER	WEIGHT
49	1	176
50	1	183
51	3.69e-40	3.69e-40

- Delete the 51st row by selecting the row (it appears with its row number reversed, as shown above).
- Choose **Rows>Delete Rows**.

● Check that the modeling types are correct. The grouping variable must have a nominal or ordinal modeling type. Notice that Gender's modeling type is Continuous; this is not correct. The modeling type of the variable of interest, Weight, should be Continuous. Weight is correctly label.

- Change the modeling type for Gender to nominal.

● Choose **Analyze>Fit Y by X**.

● Select the grouping variable Gender as the X column, and select the response variable Weight as the Y column.

A dot plot appears but because there are duplicate weights, their points appear on top of each other. To make them visible, let's wiggle them around.

● Choose **Display▶ Jitter** and a jittered dot plot appears.

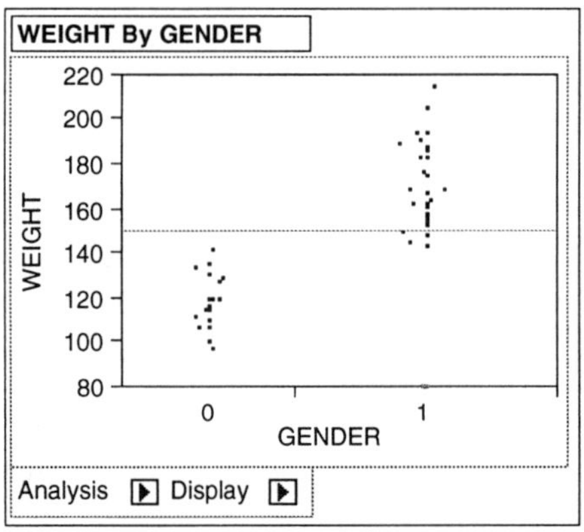

As in ES7 Figure 2-9, the display above illustrates that the Gender=0 students (the females) weigh less than the Gender=1 (the male) students.

Constructing single point plot displays

The data in Illustration 2-3 on page 42 of ES7, displayed as a dot plot in Figure 2-4 has a single group. The trick for obtaining a point plot for this data is to create an X variable with a single, constant value. Then you can apply the method above to obtain a point plot. The steps to obtain a point plot similar to Figure 2-3 are as follows.

- Import the ILL2-3.XPT data table.
  - Choose **File>Import>SAS transport file**, and
  - Select the file ILL2-3.XPT.
- Create a new column whose values are all "Group" (or some other constant with nominal modeling type). You could repeatedly type a value into this new column in all the rows but it is easier to type the value once and copy it to all rows in that column. The steps are:
  - Choose **Cols>New Column**.
  - Type **Col Name: Group**.
  - Specify **Data Type: Character** from the popup. Notice that character columns automatically have **Modeling Type: Nominal**.
  - Click **OK**.

  Notice that the data table window appears with empty rows under Group.
- Place the same value in every row of Group.
  - Type "Group" in the first row and press the Return or Enter key.
  - Highlight the word "Group" in the first row and choose **Edit>Copy**.
  - Choose **Rows>Select>All Rows** and highlight the Group column.
  - Select **Edit>Paste** to paste "Group" into all rows.

  All values in the Group column now should be "Group" (without quote marks).
- Choose **Analyze>Fit Y by X**.
- Select Group as the X column, and select EXAMGRAD as the Y column.

- Choose the **Display▶ Jitter** option and a jittered point plot appears.

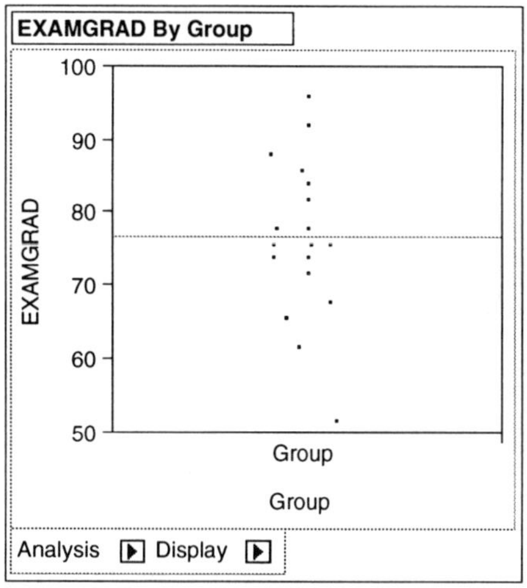

As in ES7 Figure 2-9, the display shows that most of the data is bunched around 75 and spread out at the extremes.

Real dot plots

To reproduce ES7 Figure 2-4, use the DOTPLOT JMP template. This template calculates everything you need to produce a dot plot. Complete documentation of the template is in the Appendix on page 129.

Use of the DOTPLOT template The data used in Figure 2-4 is contained in the ILL2-3 file. To create a dot plot, sort the data values, copy and paste the data values into the DotPlot template, and then launch the **Fit Y by X** platform. The detailed steps are below.

- Sort the original data table (ILL2-3) by EXAMGRAD.
  - Choose **Tables>Sort**.
  - Specify EXAMGRAD as the sorting variable.
- Copy the values of EXAMGRAD into the clipboard.
  - Highlight EXAMGRAD.
  - Choose **Edit>Copy**.
- Choose **File>Open** and select the DOTPLOT file.

The DotPlot.jmp template has two columns:

DOTPLOT

2 Cols / 0 Rows	sorted values	Frequency

- Paste the values on the clipboard into the first column.
    - Change the column name from sorted values to EXAMGRAD.
    - Highlight the first column in DOTPLOT.
    - Choose **Edit>Paste**.
- Display the dot plot.
    - Choose **Analyze>Fit Y by X**.
    - Specify Frequency as the Y column and EXAMGRAD as X.

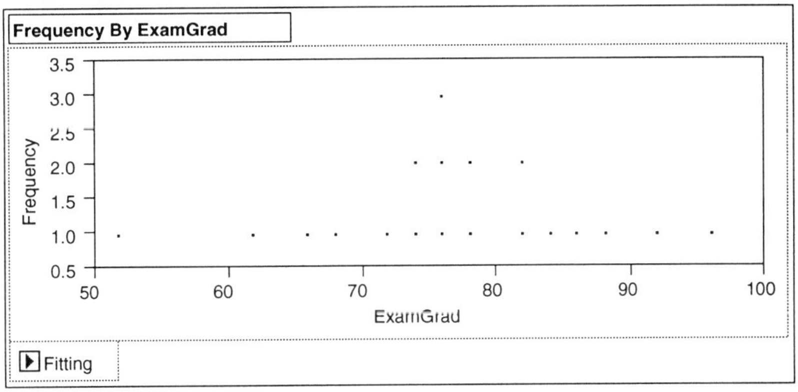

The JMP IN Fit Y by X point plot is the same as the dot plot of Figure 2-4 on ES7 pg. 43.

Constructing an ogive

An ogive is a plot of the cumulative relative proportion versus the variable of interest. The steps to create an ogive are straightforward. Generally, you simply create a variable with the cumulative proportions, and then you create the plot. Detailed instructions follow to create Figure 2-14 from the data in TAB02-5.

- Choose **Analyze>Distribution of Y**.
- Specify SCORE as the Y variable.

- Choose **Save Prob** Scores from popup menu list at the bottom left of the window.
- Change the new column name to Proportion.
- Display the Ogive.
    - Choose **Graph>Overlay Plot**.
      Note: this automatically connects the dots.
    - Specify Proportion as the Y column and SCORE as X.

The Figure 2-14 on pg. 63 of ES7 is the Ogive for 8 classes; the figure below shows the cumulative relative proportion of each value of SCORE .

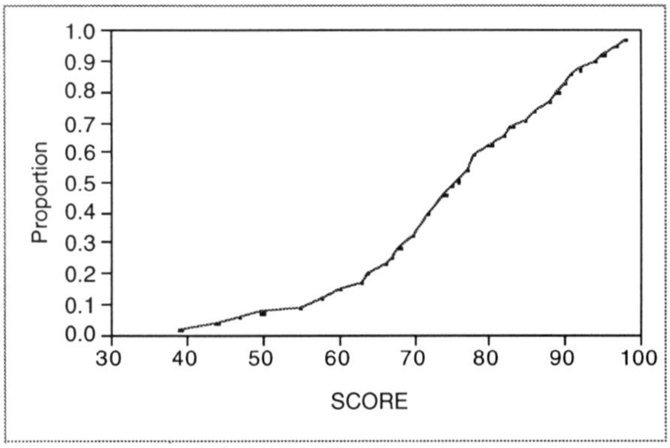

To rescale the axes, double-clicking on an axis complete the dialog that appears.

Specific Applications in ES7

This section describes the various computer applications in the second chapter of ES7. The above general information is referred to, as are some JMP templates. Templates are documented in the Appendix.

Constructing a pareto diagram (pg. 42)

JMP IN pareto diagrams are generated using the **Pareto Chart** command in the **Graph** menu. See **Pareto charts** on page 27 in this chapter for an example.

Constructing a dot plot display (pg. 43)

There are two methods for creating figures like dot plots. Point plots can be generated using the method in **Dot plots and ogives: Constructing single point plot**

displays on page 29. To create a dot plot requires the JMP calculator. This method is shown in **Dot plots and ogives: Real dot plots** on page 32.

Constructing a stem-and-leaf display (pg. 46)

Stem-and-Leaf plots are optional graphs built into the **Distribution of Y** displays. This example uses the data in TAB02-02.

- Choose **Analyze>Distribution of Y**.
- Select the Y column of interest. In Figure 2-7 the column is WEIGHT.

After the histogram appears, find the ▶ option popup at the top of the window, next to the histogram name.

- Choose **Stem and Leaf** from the popup above the histogram. The display is shown below:

Stem and Leaf

Stem	Leaf	Count
21	5	1
20	5	1
19	0155	4
18	3468	4
17	00667	5
16	122578	6
15	044578	6
14	2358	4
13	257	3
12	00089	5
11	0255688	7
10	188	3
9	8	1
8		

Multiply Stem.Leaf by 10

In JMP and JMP IN version 3.1.5 there is no provision to change the division between stem and leaf.

Constructing multiple Dot plot displays (pg. 48)

JMP IN point plots can be generated using the Fit Y by X platform. See **Dot plots and ogives: Constructing multiple point plot displays** on page 29 for an example.

Exercise 2.24 (pg. 52-53)

The stem-and-leaf plot for exercise 2.24 was produced by the following steps:

- Import EX2-24.XPT. The dataset has 25 rows.
- Choose **Analyze>Distribution of Y**.

- Select **ANNHDD>Y**.
- Choose ▶ **Stem and Leaf** from the popup above the histogram.

  The stem-and-leaf display is shown below:

Stem and Leaf

Stem	Leaf	Count
69	256	3
68	019	3
67	34	2
66	8	1
65	58	2
64	234	3
63	027	3
62	006	3
61	047	3
60	79	2

Multiply Stem.Leaf by 10^2

Notice the note below the display, "Multiply Stem.Leaf by 10^2." That is, you can reconstruct the data by forming a number composed of the stem (for example, the smallest stem=60), a decimal point (.), and the one digit leaf (the smallest leaf=7). The smallest number is thus 60.7 • 100=6070. You may also notice that because of rounding, the leaves are slightly different than the stem-and-leaf plot shown on pg. 53 of ES7.

Constructing histograms (pgs. 59-60)

Histograms are default displays whenever the **Distribution of Y** command is used. See the instructions in **Graphical displays: Histogram scaling** on page 23 for the two ways to set the number of bars displayed. To specify whether frequency counts or proportions are used to describe the length of the histogram bars, see the instructions in **Optional displays and reports: Common ☑ popup menu options** on page 26.

Constructing an ogive (pg. 63)

There are two steps to creating an ogive curve. The steps are shown in **Dot plots and ogives: Constructing an ogive** on page 33.

Finding the mean (pg. 69)
Finding the median (pg. 71)
Finding the standard deviation (pg. 82)
Finding several additional statistics (pg. 82)

All of the above statistics, plus more, are displayed by default when you use the **Distribution of Y** platform with a continuous variable. See **Using Distribution of Y to describe a variable: Text reports for continuous columns** on page 26 in this chapter.

Exercise 2.85 (pg. 85)

This exercise uses formula 2-6 to calculate the standard deviation of a column. The data is not on the student disk so you must type it in. Use the FORM2-6 template, described in the Appendix on page 130.

- Create a new data table with 10 rows.
  - Choose **File>Open**.
  - Select file FORM2-6.
  - Specify **# Rows to Add: 10**.
- Type the data given in the exercise.

  After each data value, the formulas recalculate. When the last row has been entered, the data table appears as follows.

EX2-085 using FORM2-6

6 Cols / 10 Rows	X	X bar	deviation	deviation^2	s - squared	s
1	7	7.2	-0.2	0.04	8.177778	2.859681
2	6	7.2	-1.2	1.44	8.177778	2.859681
3	10	7.2	2.8	7.84	8.177778	2.859681
4	7	7.2	-0.2	0.04	8.177778	2.859681
5	5	7.2	-2.2	4.04	0.177770	2.050681
6	9	7.2	1.8	3.24	8.177778	2.859681
7	3	7.2	-4.2	17.64	8.177778	2.859681
8	7	7.2	-0.2	0.04	8.177778	2.859681
9	5	7.2	-2.2	4.84	8.177778	2.859081
10	13	7.2	5.8	33.64	8.177778	2.859681

If you already have values in a data table, copy and paste the values into the FORM2-6 template as described in the Appendix on page 130. Note: the number of rows should match the number of data points. If there are too few or too many, you will get the wrong answer.

Exercise 2.86 (pg. 85)

This exercise uses formula 2-10 to calculate the standard deviation of a column. If you have already done exercise 2.85, just copy and paste the data into the FORM2-10 template.

- Copy the values of X into the clipboard.
  - Highlight the column X in EX2-085.
  - Choose **Edit>Copy**.
- Choose **File>Open** and select the FORM2-10 file.
- Paste the values in the clipboard into the first column.
  - Change the column name from X to the original column name from the source dataset.
  - Highlight the first column.
  - Choose **Edit>Paste**.

The calculations automatically occur and the results appears as below.

EX2-086 using FORM2-10

4 Cols / 10 Rows	X	X^2	s - squared	s
1	7	49	8.177778	2.859681
2	6	36	8.177778	2.859681
3	10	100	8.177778	2.859681
4	7	49	8.177778	2.859681
5	5	25	8.177778	2.859681
6	9	81	8.177778	2.859681
7	3	9	8.177778	2.859681
8	7	49	8.177778	2.859681
9	5	25	8.177778	2.859681
10	13	169	8.177778	2.859681

Finding the mean, variance, standard deviation of frequency distribution (pg. 89)

When using data in a frequency table, it is not necessary for you to calculate means and variances (standard deviations). All you have to do is enter the class values in one column and the frequency counts in another column. JMP

IN does all the calculations for you if you specify the Freq modeling type. For instance, the data in Exercise 2.87 (pg. 87) is shown below.

EX2-087

2 Cols / 5 Rows	x (C, Y)	f (C, F)
1	0	1
2	1	3
3	2	8
4	3	5
5	4	3

- Create the new data table and type in the values in the two columns.
- Choose **Analyze>Distribution of Y**.
- If you have not already done so, as in the data table above, Specify the column roles X **>Y>**, and F **>Freq>**.

  The histogram, box plot and **MOMENTS** report appear.
- Use the hand tool and drag the histogram bars to the right to see each of the five distinct values.

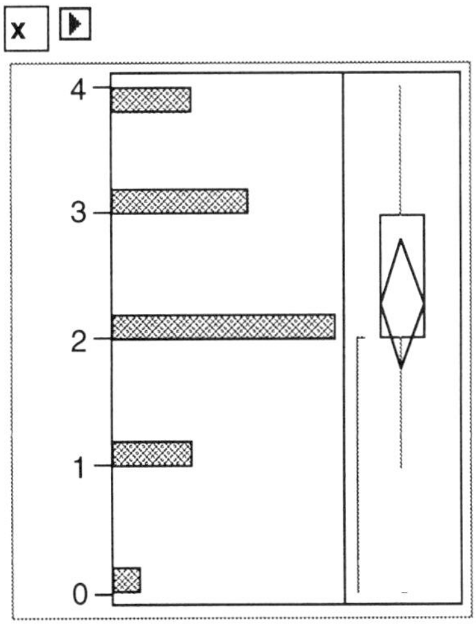

Quantiles

Moments

Mean	2.30000
Std Dev	1.08094
Std Error Mean	0.24170
Upper 95% Mean	2.80589
Lower 95% Mean	1.79411
N	20.00000
Sum Weights	20.00000

Without the use of the Freq modeling type, N is simply the number of rows in the data table. In this case, N would be 4. Marking a column as a Freq modeling type implies that you want each row to "count" for that many observation. Thus in the above example, N is the sum of the values of the Freq variable.

Finding the "5-number summary" (pg. 99)

The summary statistics used in the box-and-whisker plot are displayed in the **Quantiles** text report by default when you use the **Distribution of Y** platform with a continuous variable. See **Using Distribution of Y to describe a variable: Text reports for continuous columns** on page 26 in this chapter.

Constructing box plots (pg. 100)

An outlier box plot is one of the two default displays for a continuous variable in the **Distribution of Y** platform. See the instructions in **Graphical displays** on page 23.

Some data management commands: Sort, Group/Summary, and Print (pg. 101)

The section **JMP Data Tables** in Chapter 1 shows how to modify JMP data tables (see page 10). On page 21, the explanation of how to print your results or how to save results in a word processing file is presented.

Generating random numbers (pg. 101-102)

JMP calculator formulas can be used to generate random numbers. Random number generators for the following distributions are available: Uniform, normal, single parameter exponential, Cauchy, gamma, , triangular, a random permutation of the row numbers, Poisson, binomial, geometric, and negative binomial. See your documentation for details.

JMP calculator Random Number Functions

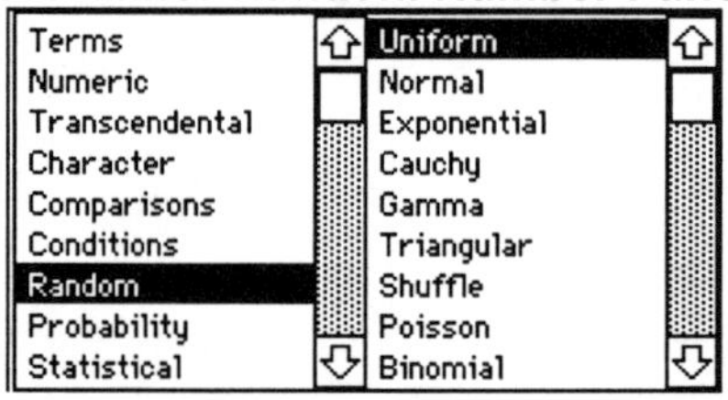

In addition, general purpose JMP templates are provided in the Appendix to generate specific distributions. Templates are provided for the distributions discussed on the bottom of ES7 pg. 101, and others. Since JMP IN uses the calculator to generate random variables, different random values can be recalculated by pressing the **Evaluate** button in the JMP IN calculator.

Selecting a sample without and without replacement (pg. 101)

Most of the Random Number Generation Templates, in the Appendix beginning on page 141 generate samples from continuous distributions and thus the with and without distinction is not important. Those that generate discrete values (**RANCOIN**, **RANDIE6**, **RANINT**, and **RANDISC**) effectively sample with replacement. To create a subset of observations without replacement from a fixed set of numbers, see the Exercise 2.188, below.

Normal probability paper and testing for normality (pg. 107)

Figure 2-35 in ES7 shows an ogive plotted on probability paper. The vertical scale is the cumulative normal probability. This display is an optional display of the **Distribution of Y** platform. The data in TAB02-5 is used in the example on page 44.

- Choose **Analyze>Distribution of Y**.
- Specify **SCORE** as the Y variable.
- Choose ☑ **Normal Quantile Plot** from the popup menu at the bottom left border of the window. The normal quantile plot using all of the continuous values of SCORE is added to the histogram and box plot.

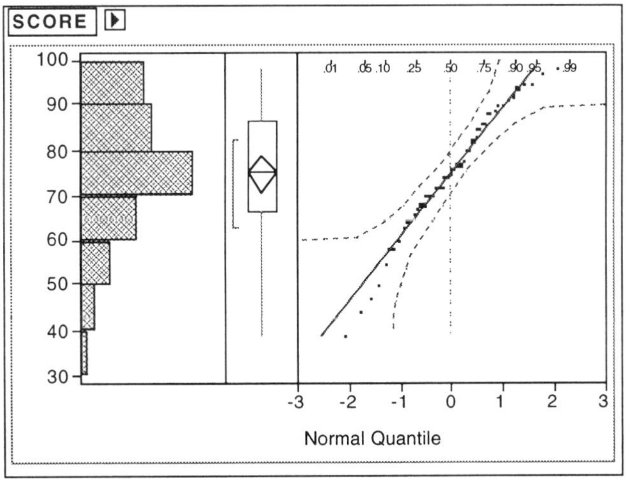

Notice that two horizontal scales are plotted on the Normal Quantile plot: the cumulative relative frequency proportions (on the upper horizontal axis) and the z-score (on the lower horizontal axis). To graphically assess normality, check if all of the data points are within the 95% Lillifors confidence bounds. See JMP help for a discussion of these bounds.

Other tests of normality include the empirical rule discussed in ES7 pgs. 105-106. The JMP template NORMRULE and **Tables>Group/Summary** was used to calculate the following table of observed proportions.

SCORE from TAB2-05

3 Cols / 1 Rows	Mean(~68%)	Mean(~95%)	Mean(~99.7%)
1	0.66	0.96	1

See the documentation of the NORMRULE template in the Appendix.

Exercise 2.137 (pg. 109)

In this exercise use the RANNOR template to generate K=100 samples (T=1 trials) from a normal distribution with mean 50 and standard deviation 10.

- Use the RANNOR template according to the instructions in the Appendix on page 142. A portion of one set of data is shown below.

EX2-137 using RANNOR

4 Cols / 100 Rows	# samples per trial	Trial	Sample	Normal (50, 10)
1	100	1	1	43.26092
2	•	1	2	39.96981
3	•	1	3	62.38837

The histogram, box plot, and normal quantile plot do not indicate any departures from normality.

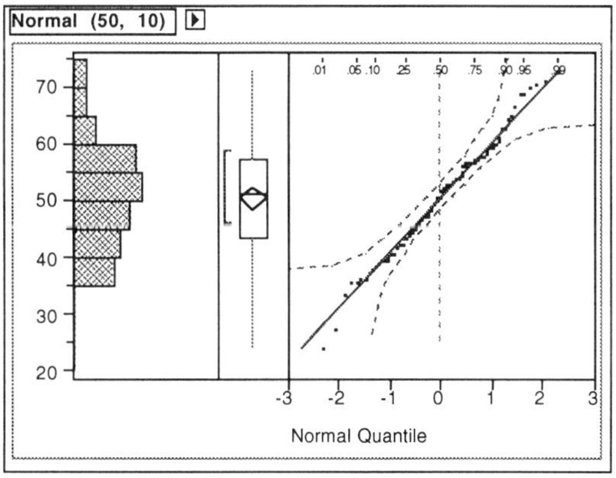

The empirical rule can be calculated using the NORMRULE template. It calculates the observed proportion within 1, 2, and 3 standard deviations.

- Use the NORMRULE template and **Tables>Group/Summary**. The observed mean proportions for one set of 100 samples is shown below.

3 Cols / 1 Rows	C Y Mean(~68%)	C Y Mean(~95%)	C Y Mean(~99.7%)
1	0.68	0.94	1

Part b of exercise 2.137 asks you to repeat the above calculation with a different sample.

- Double-click the column name Normal(50,10) in the original data table.
- Press the mini formula display button at the bottom of the **Col Info** dialog.
- Press the [Evaluate] button in the calculator window.
- Close the calculator window.
- Use the NORMRULE template and **Tables>Group/Summary** again.

Exercise 2.138 (pg. 110)

In this exercise use the RANEXP template to generate K=100 samples (T=1 trials) from a exponential distribution with mean/standard deviation 50.

- Use the RANEXP template according to the instructions in the Appendix on page 147. A portion of one set of data is shown below.

EX2-138 using RANEXP

4 Cols / 100 Rows	C # samples per trial	N Trial	N Sample	C Y Exponential (50)
1	100	1	1	34.16833
2	•	1	2	27.70632
3	•	1	3	43.33688

The histogram, box plot, and normal quantile plot indicate non-normal data.

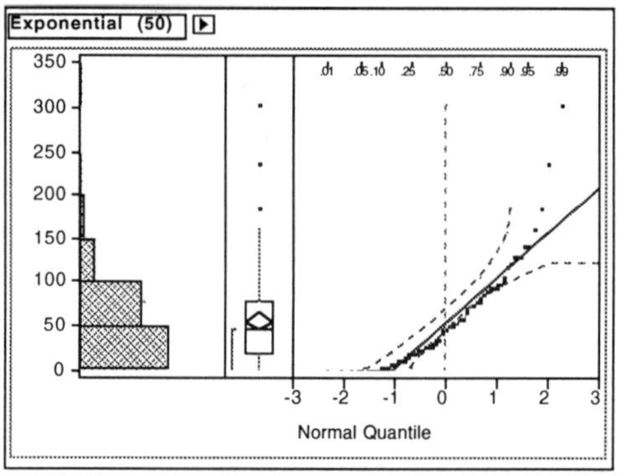

- Use the **NORMRULE** template, as in exercise 2.137 above.

3 Cols / 1 Rows	C Y Mean(~68%)	C Y Mean(~95%)	C Y Mean(~99.7%)
1	0.78	0.96	0.98

Exercise 2.188 (pg. 127)

- Use the **RANNOR** template to generate K=500 samples in a single trial of random values with mean=100 and standard deviation of 20.

  The easiest way to extract a subset of 30 of these rows is with the calculator.

- Add a new, calculated column called **Subset 1** with only 30 of the original 500 values.
  - Choose **Cols>New Column**.
  - Type **Col Name: Subset 1**.
  - Specify **Data Source: Formula** from the popup.
  - Click **OK**.
  - Create the following formula.

$$\begin{cases} \textit{Normal (100, 20)}, & \text{if ?shuffle} \leq 30 \\ \square, & \text{otherwise} \end{cases}$$

The formula uses the Conditions>IF function, the Comparisons> x≤y function, and the Random>Shuffle function. Shuffle randomly permutes the row numbers yielding a number between 1 and 500, in this case. Thus, values less than or equal to 30 indicate a random subset of the 500 rows.

- Add another calculated column called Subset 2 with only 30 of the original 500 values. Use copy and paste to reuse the formula you built for Subset 2.
  - Click on the entire formula for Subset 1 while in the calculator window.
  - Choose **Edit>Copy**.
  - Choose **Cols>New Column**.
  - Type **Col Name: Subset 2**.
  - Specify **Data Source: Formula** from the popup.
  - Click **OK**.
  - Choose **Edit>Paste**.
- Add another calculated column called Subset 3 using copy and paste, as above.
- Add another calculated column called Subset 4 using copy and paste, as above. A portion of one set of data is shown below.

EX2-188

8 Cols / 500 Rows	C Y Normal (100, 20)	C Y Subset 1	C Y Subset 2	C Y Subset 3	C Y Subset 4
31	120.1	120.1	•	•	•
32	90.9	•	•	•	90.9
33	88.7	•	88.7	•	•
34	86.1	•	•	•	•
35	89.0	•	•	•	•

To compare the population data in the column Normal (100,20) and the three subsets, use the **Distribution of Y** platform.

- Choose **Analyze>Distribution of Y**.
- Select the following Y columns: Normal (100,20), Subset 1, Subset 2, Subset 3, and Subset 4.
- Choose ☑ **Uniform Axes** from the popup menu at the bottom left border of the window. This forces all of the histograms on the same scale. The population and one subset are shown below.

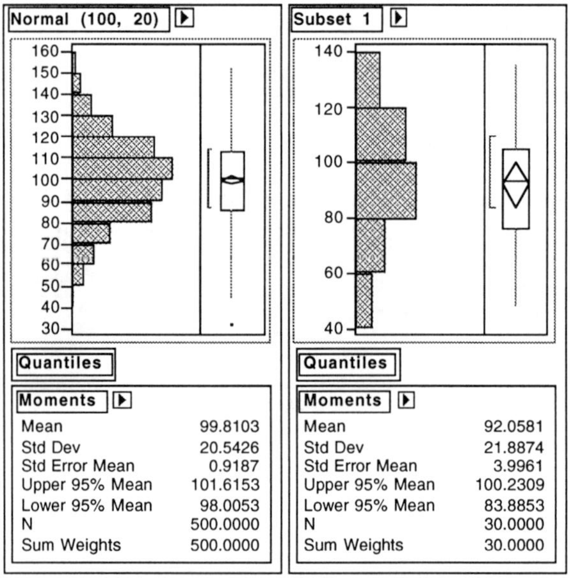

Notice that N is 500 in the column with the population data and N is 30 in the subset columns. The graphical displays and numerical summaries allow you to compare the population and random subsets of various sizes.

Chapter 3
Bivariate Data

Fit Y by X

Bivariate data is data in two different columns obtained from the same observation. Most commonly, JMP IN looks at relationships between pairs of columns using the Fit Y by X platform in the Analyze menu.

Data table layout

When working with bivariate data with JMP IN the two variables must be in separate columns. Observations in the same row are assumed to come from the same case (e.g. the same sampling unit, the same subject). The values in the columns describe the characteristics of the row. So, throughout this chapter, you must organize the data into (at least) two columns. If you also have frequency count data, be sure to add this column of information to the data table and use the role assignment popup menu to mark it with the Freq role.

Roles and modeling type

The Y role is the response variable role and the X role is the explanatory variable role. In *some* cases, reversing the Y and X role is OK, but in most cases it is not. Recall also that columns also have modeling types of continuous, ordinal or nominal. Modeling type is a key JMP concept; the modeling type you choose determines how JMP IN uses the data in a variable. ES7 uses the term *quantitative* to refer to continuous columns and *qualitative* to refer to one of the two categorical modeling types. Since a Y column can be either continuous or categorical and an X column can be either continuous or categorical, there are four situations you can encounter when looking at bivariate data.

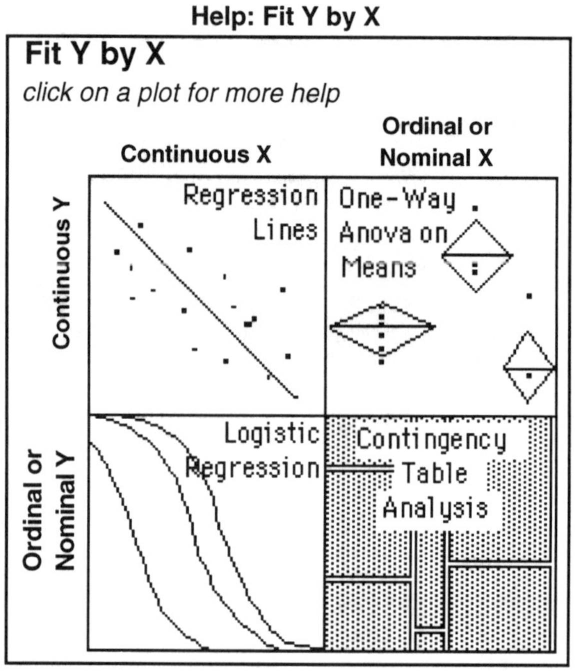

Two qualitative variables

ES7 Chapter 3 refers first to the situation where both columns are qualitative. In JMP terminology, both variables are categorical: nominal or ordinal modeling type. In the above figure this is illustrated by a contingency table analysis. Refer to the cross-tabulation analysis is shown in the Specific Applications section below. See chapter 9 in JSS for more discussion of categorical distributions.

One qualitative and one quantitative

The next section of Chapter 3, ES7 refers to the bivariate data situation where one column is qualitative and the other is quantitative. ES7 covers the situation where the response variable is continuous (quantitative) and the explanatory variable is categorical (qualitative). This is the JMP one-way layout. Refer to the section on Comparing two sets of data in the Specific Application section below. Also see chapter 6 in JSS for test of differences between two means. The other possibility, called logistic regression in the

Figure above, is probably too advanced for an elementary text and is not covered here.

Two quantitative variables

Most of ES7 focuses on the situation where both variables are quantitative. In JMP this corresponds to both Y and X being continuous. Various forms of regression and curve fitting are available. A number of Specific Applications of this are shown below. JSS chapter 8 discusses fitting curves through points.

Specific Applications in ES7

Cross-tabulation tables (pg. 140)

Cross-tabulation tables are automatically displayed in the Fit Y by X platform when both columns are categorical, either nominal or ordinal modeling type. See the data in exercise 3.9 (ES7 pg. 145).

Modeling Type In the data file PURCHASE is coded 1=>Yes and 2=>No. Gender is coded 1=>M and 2=>F. Since both variables are numeric they automatically receive the continuous modeling type. Change modeling type to nominal.

Column Roles In JMP IN a response variable is given the Y role and the explanatory variable is the X role. Is PURCHASE the response column or is GENDER the response column? In this case it is clear that purchasing an item is the response that can be influenced by the gender of the customer. However, in this case it does not really matter which you chose as X or Y. In contingency-tables the choice of X and Y affects which column is used for rows in the resulting contingency-table (the Y variable) and which is used as columns (the X variable).

- Import the data table.
  - Choose **File>Import>SAS transport file**.
  - Select the file TAB3-09.XPT. When you inspect the data, you will see that the last row must be deleted.
  - Select the 61st row by clicking.
  - Choose **Rows>Delete Rows**.

- Check that the modeling types are correct. Both grouping variables should have a nominal modeling type.
    - Change the modeling type for **PURCHASE** to nominal (N).
    - Change the modeling type for **GENDER** to nominal (N).
- Choose **Analyze>Fit Y by X**.
- Select one variable to be Y and the other to be X.
    - Select PURCHASE **>Y>**.
    - Select GENDER **>X>**.
    - If you also have a frequency count variable, select it as Freq.

A *mosaic chart*, **Crosstabs** report, and **Tests** report appear. You need not concern yourself with the statistical tests until Chapter 10 so we'll ignore them for now. Remove the statistical tests report by unchecking ▶ **Tests** in the popup at the top of the window.

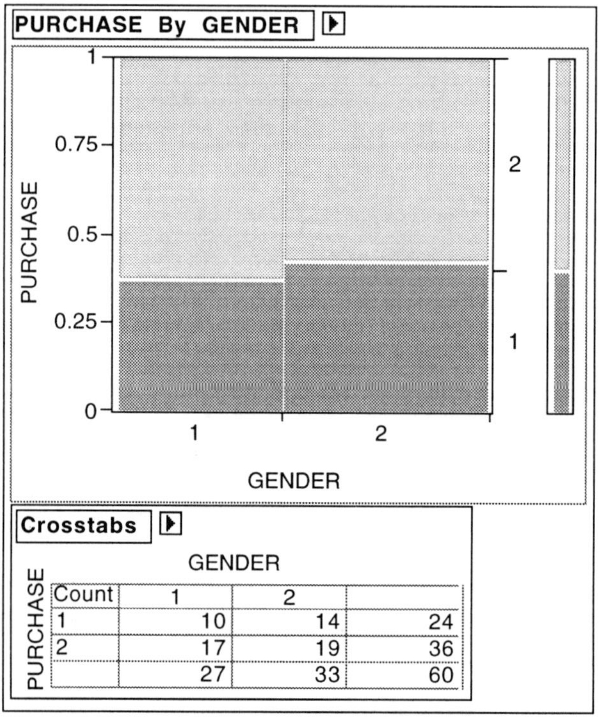

Crosstabs

PURCHASE / Count	GENDER 1	GENDER 2	
1	10	14	24
2	17	19	36
	27	33	60

If you see a scatterplot with just a few points rather than a mosaic plot, you forgot to set the modeling types to nominal.

Interpreting the mosaic chart Use the help tool (**?**) to assist you in the interpretation of the mosaic chart. In all JMP IN plots the Y axis is vertical and the X axis is horizontal. The two genders appear across the figure in different stacked bars. The two classes of purchasers appear as different colors within the gender bars since JMP IN plotted the Y variable on the vertical axis. In this case, the lower red bar is the proportion of one gender that was PURCHASE =1. Since the red bar is shorter for GENDER =1 than for GENDER =2, there are less purchasers (1) in the male group (1) than the female group (2).

Notice that the **CROSSTABS** report has an option popup menu. The default value in the table is the Count. There are 10 rows in the data table with PURCHASE=1 and GENDER =1. The popup enables you to choose ▶ **Total %** if you want to display the counts÷grand total as a percent, or to choose ▶ **Row %** if you want to display the counts÷row total as a percent, , or to choose ▶ **Col %** if you want to display the counts÷column total as a percent. If all three choices are made, the following **CROSSTABS** report results.

Crosstabs ▶

	Count Total % Row % Col %	GENDER 1	2	
PURCHASE	1	10 16.67 41.67 37.04	14 23.33 58.33 42.42	24 40.00
	2	17 28.33 47.22 62.96	19 31.67 52.78 57.58	36 60.00
		27 45.00	33 55.00	60

In the first cell (PURCHASE =1 and GENDER =1) The percentages are calculated as follows:

Total % = 10/60 = .1667 = 16.67%,

Row % = 10/24 = .4167 = 41.67%, and

Col % = 10/27 = .3704 = 37.04%.
Grammatically, one way to state these results is, "Of all subjects surveyed 16.67% were male and purchased. Of all purchasers 41.67% were male. Of all males 37.04% were purchasers." It is the denominator which makes the difference in the interpretation.

Comparing two sets of data (pg. 143)

Construction of box plots or dot plots in ES7 corresponds to the bivariate situation where one column is quantitative and one is qualitative. When two or more sets of rows identified by an X grouping variable are compared on a Y response variable in JMP IN using the Fit Y by X platform, the default display is a point plot. Point plots to compare groups were discussed in **Constructing multiple point plot displays** on page 29. Box plots are an optional display. To continue with the data in exercise 3.9, let's compare the ages of the purchasers and non-purchasers.

Column Roles When comparing two groups, always mark the quantitative variable as [Y] and the qualitative variable as [X]. The X groups are compared on their values of the Y column.

Modeling Type In the original data file PURCHASE is coded 1=>Yes and 2=>No and AGE is numeric. Since both variables are numeric they automatically received the [C]ontinuous modeling type. Change modeling type to [N]ominal for PURCHASE since you want it to identify the two different groups.

- Import the data table. In this example, use the TAB3-09 data in the previous example.
- Check that the modeling types are correct. The grouping variable should have a nominal modeling type.
  - Verify the modeling type for AGE is [C]ontinuous.
  - Change the modeling type for PURCHASE to [N]ominal.
- Choose **Analyze>Fit Y by X**.
- If you have not specified roles, a role assignment dialog appears.
  - Select PURCHASE [>X>].
  - Select AGE [>Y>]. A point plot appears.

- Choose the **Display▶ Jitter** option and the display changes to a jittered point plot.
- Choose the **Analysis▶ Quantile** option and a box plot and **Quantiles** report is added.

AGE By PURCHASE

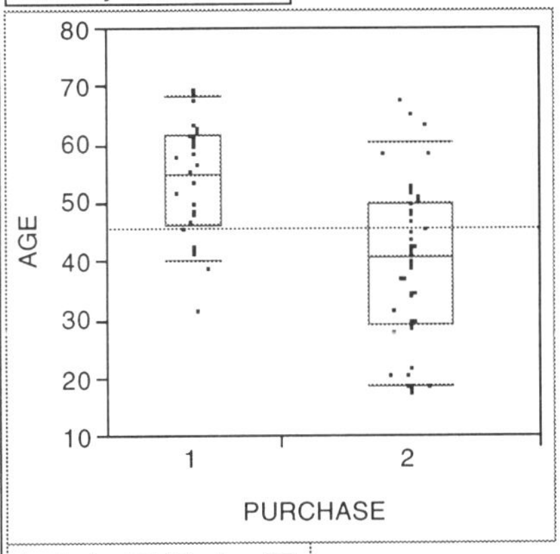

Analysis ▶ Display ▶

Quantiles

Level	minimum	10.0%	25.0%	median	75.0%	90.0%	maximum
1	32	40.5	46.25	55	61.75	68.5	70
2	18	19	29.25	41	50	60.5	68

If you want to also see the means and standard deviations for each group,

- Choose the **Analysis▶ Means, Std Dev, Std Err** option. Error bars are added to the plot and a **MEANS AND STD DEVIATIONS** report appears.

Means and Std Deviations ▶

Level	Number	Mean	Std Dev	Std Err Mean
1	24	53.9583	9.9847	2.0381
2	36	40.3889	14.1911	2.3652

The displays and reports indicate the typical values for each group as well as the range of values.

Constructing scatter diagrams (pg. 144)

The scatter plot of physical fitness data in ES7 Figure 3-5 comes from the data in TAB3-10. The scatter diagram is the default display when both columns are continuous and the Fit Y by X platform is used.

Modeling Type When variables are quantitative and do not indicate groups, use the Continuous modeling type. In the data file PUSH_UPS and SIT_UPS are both numeric. They automatically receive the continuous modeling type.

Column Roles When both columns are continuous, the Fit Y by X platform plots the two sets of values with Y on the vertical axis and X on the horizontal axis. In Chapter 3 there are two situations: 1) you are looking for a *correlation* between two columns, or 2) you are doing *regression* and trying to predict one variable from another. In situation 1) it does not matter which column you choose as X or Y. In situation 2) choose the predictor variable as X and the response variable as Y.

- Import the data table.
  - Choose **File>Import>SAS transport file.**
  - Select the file TAB3-10.XPT. When you inspect the data you will see that the last row must be deleted.
  - Select the 11th row by clicking.
  - Choose **Rows>Delete Rows.**
- Check that the modeling types are correct. Both variables should have a continuous modeling type.
- Choose **Analyze>Fit Y by X.**
- Select one variable to be Y and the other to be X. In Figure 3-5 push-ups are on the horizontal axis and sit-ups are on the vertical axis.
  - Select SIT_UPS as Y and Select PUSH_UPS as X.

A scatter diagram similar to Figure 3-5 appears.

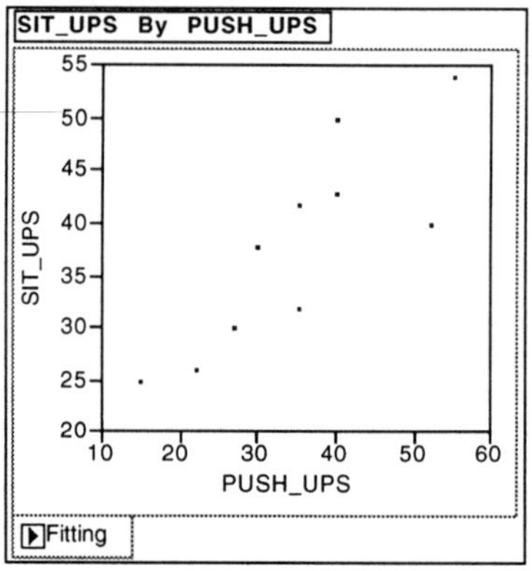

Finding the correlation coefficient (pg. 154-155)

In the example above, the correlation between the two columns are obtained by an option in the Fitting popup menu below the scatter plot.

- Choose the **Fitting▶ Density Ellipses>.95** option and a 95% confidence ellipse appears on the plot and the **Bivariate** report is added.
- Press the **Bivariate** report button and the correlation is displayed.

Bivariate

Variable	Mean	Std Dev	Correlation	Signif. Prob	Number
PUSH_UPS	35.1	12.45838	0.839439	0.0024	10
SIT_UPS	38	9.763879			

If you have more than two variables and want to see the correlations between them all, use **Analyze>Correlations of Y's**. See page 125 for an example.

If you need to see how to calculate the correlation as in Table 3-11 using equations 3-2 through 3-4, use the FORM3-2 template.

TAB3-11 using FORM3-2

6 Cols / 10 Rows	X	Y	X^2	Y^2	X•Y	r
1	27	30	729	900	810	0.84
2	22	26	484	676	572	0.84
3	15	25	225	625	375	0.84
4	35	42	1225	1764	1470	0.84
5	30	38	900	1444	1140	0.84
6	52	40	2704	1600	2080	0.84
7	35	32	1225	1024	1120	0.84
8	55	54	3025	2916	2970	0.84
9	40	50	1600	2500	2000	0.84
10	40	43	1600	1849	1720	0.84

The FORM3-2 template is documented in the Appendix on page 133.

Finding the regression equation (pg. 167-168)

Fitting a linear regression is an optional command in the Fit Y by X platform. Figure 3-21 shows the line of best fit for the data in TAB3-10. The data and related analyses are shown in the previous two examples. To add a regression line, do the following.

- Choose the **Fitting▶ Fit Line** option from the popup menu below the scatter plot.

SIT_UPS By PUSH_UPS

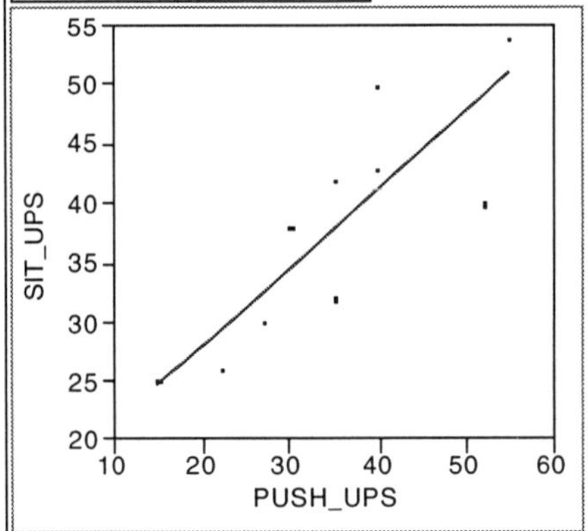

▶ Fitting ▶ —— Linear Fit

Linear Fit

SIT_UPS = 14.9082 + 0.65789 PUSH_UPS

Summary of Fit

RSquare	0.704658
RSquare Adj	0.66774
Root Mean Square Error	5.628092
Mean of Response	38
Observations (or Sum Wgts)	10

Analysis of Variance

Source	DF	Sum of Squares	Mean Square	F Ratio
Model	1	604.59661	604.597	19.0872
Error	8	253.40339	31.675	**Prob>F**
C Total	9	858.00000		0.0024

Parameter Estimates ▶

Term	Estimate	Std Error	t Ratio	Prob>|t|
Intercept	14.908225	5.577096	2.67	0.0282
PUSH_UPS	0.6578853	0.150584	4.37	0.0024

The equation for the line is given at the top of the **Linear Fit** report.

Fitting prediction equations other than a straight line (pg. 160)

Many times a line does not fit the data well. On page 160 of ES7 some other models of prediction equations are given. All of these, plus many more are available in the **Fitting▶ Polynomial** or the **Fitting▶ Transformed** popup below the scatter diagram. Some of these are covered in Chapter 13 of ES7.

Chapters 4 and 5
Probability and Random Variables

Random numbers

Chapters 4 and 5 in ES7 discuss probability and probability distributions. One way to explore what all of this means in practice is to actually do an experiment like tossing a coin a large number of times and see whether half come up heads. As you saw on page 187, Karl Pearson tossed a coin 24,000 times. Yikes! That must have been tedious. Well, fortunately there is another way to do the same sort of thing; do a simulation. It turns out that it is pretty easy to tell JMP IN to act like it is tossing 24,000 coins (or even a million coins). The computer can also easily count up how many came up heads. In this chapter you see how to do these sorts of simulations. The idea is, once you see what sorts of things happen by "dumb luck" you are better able to judge how remarkable a real life event is. For instance, in a recent *Dilbert* cartoon, the boss remarked, "You guys can't fool me. I did a study and 40% of your 'sick' days were on Mondays and Fridays!"

Random functions in the calculator

There is a whole set of functions to generate random values in the JMP IN calculator. In JSS in the section on *Random Number Functions* in chapter 3 it shows how to build an example random number generator. More complicated distributions are documented in JSS in Appendix C. Using these methods you can certainly build your own but in this companion we have tried to make it easier. There are a number of JMP templates in the **Appendix: Random Number Generation Templates** starting on page 142. They are all discussed in this and other chapters in this companion.

Plotting a series of trials

One very useful way to see what may be going on in a set of data is shown in Figure 4-2 in ES7. A simple plot of the data values across time can be very informative. If you have a column of numeric values in JMP IN, the simplest way to accomplish this is to use the Overlay Plots platform in the Graph menu. For instance, if you had data such as that in Table 4-2 the following steps create a graph like Figure 4-2.

- Choose **Graph>Overlay Plots**.
- Specify just a Y column. If you do not specify an X column, JMP IN simply plots the Y values versus the row number.
- To add a reference line, double-click on the vertical axis and choose the **Reference Lines>Add Ref. Line** option.

Figure 4–2a

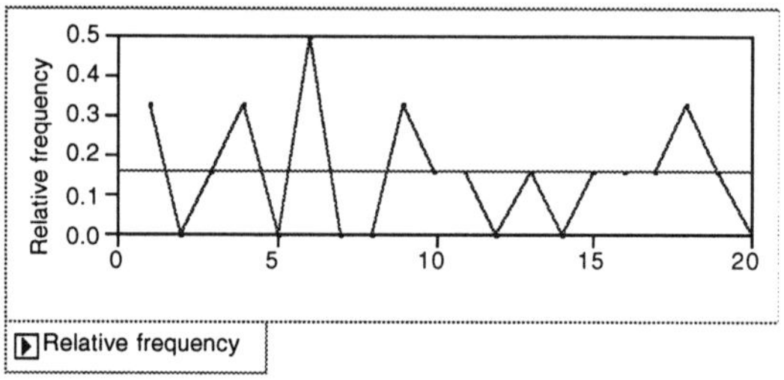

Figure 4–2b

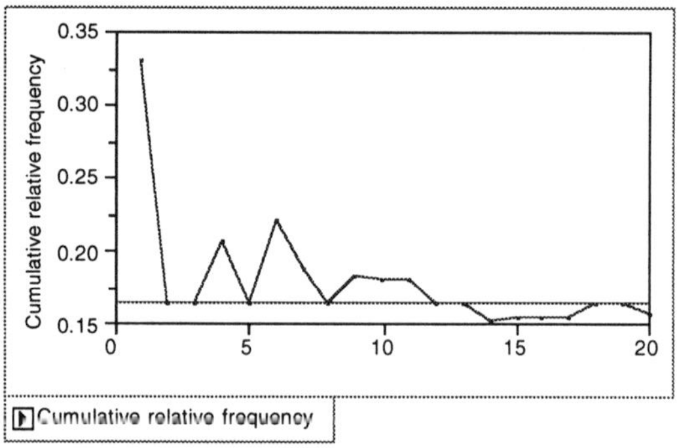

The figures do not match those in ES7 because a different set of simulated values was used.

Specific Applications in ES7 in Chapter 4

Generating random 0/1 integers (pg. 192)

On page 190 of ES7 an experiment of tossing two coins ten times is discussed. The RANCOIN template can be used to simulate coin tosses. The template is set up to toss a single coin so let's start there. For example, if there are K=10 samples in a single coin experiment, the results follow.

RANCOIN

4 Cols / 10 Rows	# samples per trial	Trial	Sample	Coin 0/1
1	10	1	1	0
2	•	1	2	0
3	•	1	3	1
4	•	1	4	0
5	•	1	5	0
6	•	1	6	1
7	•	1	7	1
8	•	1	8	1
9	•	1	9	1
10	•	1	10	0

See the documentation in **Appendix: RANCOIN** page 144 for instructions on the use of the RANCOIN template. To simulate two coins instead of one, add a new column with the same formula as is used in Coin 0/1.

- Double-click above the column name Coin 0/1.

  The Column Information dialog appears. Change the column name to Coin 1.

- Click the mini-formula button in the lower left corner to display the formula in the JMP calculator.
- Copy the formula into the clipboard: Select the formula by clicking. Choose **Edit>Copy**. Close the calculator window.
- Create a new column called Coin 2 with the same formula: Choose **Cols>New Column**. Specify **Col name: Coin 2**. Specify **Data Source: Formula**. Press **OK**. The calculator appears. Choose **Edit>Paste**. The same formula appears. Close the calculator window.

  Now there are two coins, each coming up with 0's and 1's.

In the first example of section 4.1 in ES7, the experiment is to count the number of heads in two coins. So, create a new column adding the number of heads (1) in two coins.

- Create a new column called # Heads with a formula: Choose **Cols>New Column**. Specify **Col name: # Heads**. Specify **Data Source: Formula**. Press **OK**. The calculator appears. Build the following formula and then close the calculator window.

$$Coin\ 1 + Coin\ 2$$

The top few rows of the data table appear as below (although you will have different results since you are simulating your own coin tosses).

TAB04-01 using RANCOIN

6 Cols / 10 Rows	C # samples per trial	N Trial	N Sample	C Coin 1	C Coin 2	C # Heads
1	10	1	1	0	0	0
2	•	1	2	0	1	1
3	•	1	3	1	1	2
4	•	1	4	0	1	1
5	•	1	5	0	0	0

To build your own version of Table 4-1 all you need to do is add rows corresponding to more trials. In the present table there is one trial with 10 samples (10 pairs of coins). In Table 4-1 there are 19 more trials, each with 10 samples. If you started with 10 rows, as above, you need to add 190 more.

- Choose **Rows>Add Rows**. Specify **# of Rows to add: 190**.

  Table 4-1 is simply a cross-tabulation of the number of heads versus the trial number. Use the method in **Cross-tabulation tables** on page 49 to create a crosstabs report

- Change the modeling type of # Heads to Nominal. Choose **Analyze>Fit Y by X**. Specify # Heads as Y and Trials as X. A portion of the **Crosstabs** report appears below.

 # Heads By Trial

Crosstabs

# Heads	Count	1	2	3	4	5	6
	0	2	0	2	3	5	2
	1	5	6	5	3	3	5
	2	3	4	3	4	2	3
		10	10	10	10	10	10

Figure 4-1 is the relative frequency histogram for # Heads that can be generated by the **Distribution of Y** platform.

 # Heads

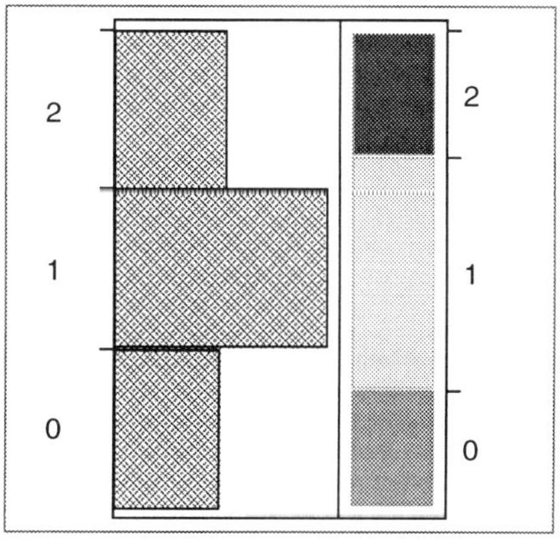

Frequencies

Level	Count	Probability	Cum Prob
0	50	0.25000	0.25000
1	98	0.49000	0.74000
2	52	0.26000	1.00000
Total	200		
	3 Levels		

Exercise 4.14 (pg. 197)

In the above example you flipped two coins. In exercise 4.14 you roll two die. Use the **RANDIE6** template, documented in the Appendix on page 146 and the method in the example above to create the following data table.

EX4-14 using RANDIE6

6 Cols / 100 Rows	[C] # samples per trial	[N] Trial	[N] Sample	[N] Die 1	[N] Die 2	[N] Sum
1	100	1	1	3	4	7
2	•	1	2	2	6	8
3	•	1	3	4	2	6
4	•	1	4	3	5	8
5	•	1	5	1	4	5
6	•	1	6	5	6	11

Note that the die and sum columns all have nominal modeling type. A **Distribution of Y** report on the three columns, Die 1, Die 2, and Sum for 100 rolls appears below.

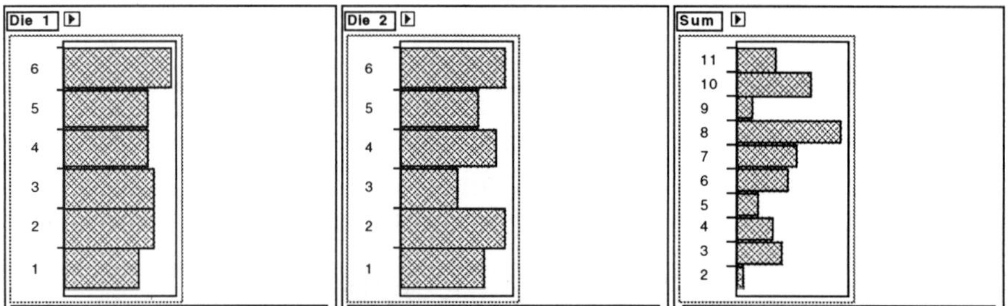

Frequencies (Die 1)

Level	Count	Probability	Cum Prob
1	14	0.14000	0.14000
2	17	0.17000	0.31000
3	17	0.17000	0.48000
4	16	0.16000	0.64000
5	16	0.16000	0.80000
6	20	0.20000	1.00000
Total	100		
	6 Levels		

Frequencies (Die 2)

Level	Count	Probability	Cum Prob
1	16	0.16000	0.16000
2	20	0.20000	0.36000
3	11	0.11000	0.47000
4	18	0.18000	0.65000
5	15	0.15000	0.80000
6	20	0.20000	1.00000
Total	100		
	6 Levels		

Frequencies (Sum)

Level	Count	Probability	Cum Prob
2	2	0.02000	0.02000
3	10	0.10000	0.12000
4	8	0.08000	0.20000
5	5	0.05000	0.25000
6	11	0.11000	0.36000
7	13	0.13000	0.49000
8	22	0.22000	0.71000
9	4	0.04000	0.75000
10	16	0.16000	0.91000
11	9	0.09000	1.00000
Total	100		
	10 Levels		

The mosaic plots in the above figure were removed by unchecking **Mosaic Plot**.

Exercise 4.29 (pg. 203)

The methods in the two examples above can be combined to solve exercise 4.29. A portion of an example data table is shown below. Use the formula in RANCOIN template for variable Coin and use the formula in RANDIE6 template for variable Die.

EX4-29

5 Cols / 200 Rows	# samples per trial	Trial	Sample	Coin	Die
1	200	1	1	0	4
2	•	1	2	0	6
3	•	1	3	0	2
4	•	1	4	0	5
5	•	1	5	1	4
6	•	1	6	1	6

The combined sample space for the above realization of the experiment can be summarized in the following Crosstabs report from Fit Y by X.

Coin By Die

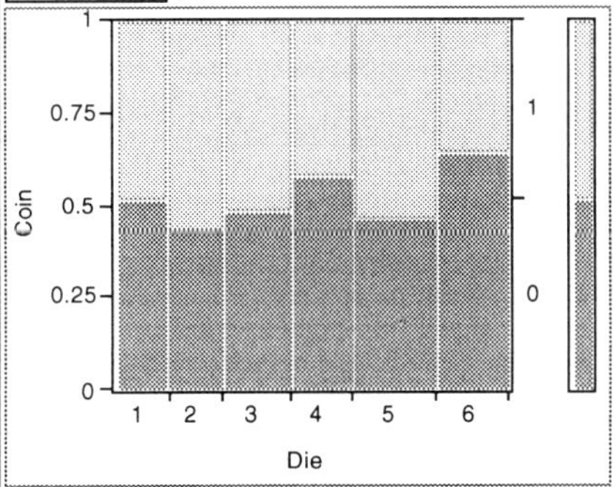

Crosstabs

Die

Coin	Count	1	2	3	4	5	6	
	0	14	12	17	18	20	23	104
	1	13	16	18	13	23	13	96
		27	28	35	31	43	36	200

Specific Applications in ES7 in Chapter 5

Generating random data with specific outcome probabilities (pg. 256)

The RANDISC JMP template generates random integers with specific probabilities. The probabilities are entered into one column and the randomly generated integers are calculated. See exercise 5.19 for an example.

Exercise 5.19 (pg. 259)

This exercise generates a random sample of 25 observations from a specified discrete probability distribution. Use the RANDISC template documented in the Appendix on page 148 for this exercise. After opening the template, the steps for completing exercise are as follows.

- Type the probabilities in the first 5 rows of Prob.
- Type 25 into the column # samples per trial.
- Add as many rows as are required to total K=25. A portion of the final data table appears below.

EX5-019 using RANDISC

5 Cols / 25 Rows	Prob	# samples per trial	Trial	Sample	Discrete
1	0.2	25	1	1	2
2	0.3	•	1	2	5
3	0.3	•	1	3	2
4	0.1	•	1	4	3
5	0.1	•	1	5	3
6	•	•	1	6	2
7	•	•	1	7	1

- Choose **Analyze>Discribution of Y**.
- Specify that the Y variable is Discrete.

Discrete

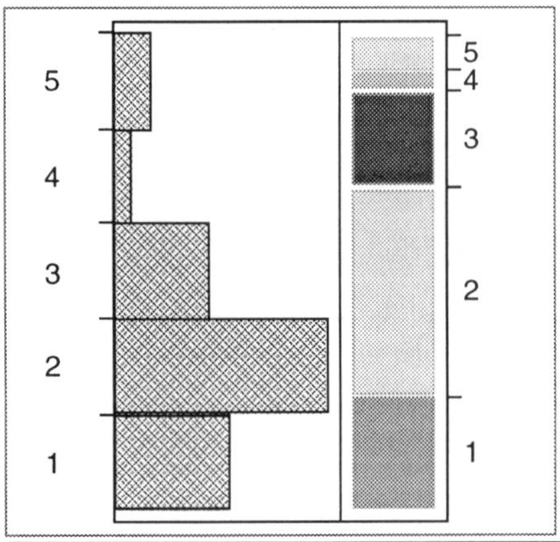

Frequencies

Level	Count	Probability	Cum Prob
1	6	0.24000	0.24000
2	11	0.44000	0.68000
3	5	0.20000	0.88000
4	1	0.04000	0.92000
5	2	0.08000	1.00000
Total	25		
	5 Levels		

To generate another set of 25 values, type over any of the values in Prob. When values in Prob change (even when they change to the same value), JMP IN recalculates another set of random values for Discrete.

Calculating binomial pdf and cdf values (pg. 273)

To determine the binomial probabilities (pdf) and binomial cumulative probabilities (cdf) for a specified binomial probability distribution, use the BINOMIAL template documented on page 150 in **Appendix: BINOMIAL**. For an example of its use, see exercises 5.73 and 5.76 below.

Exercise 5.73 (pg. 276)

This exercise calculates the binomial probabilities for n=30 and p= 0.35. The steps in using the BINOMIAL template in this case are given below.

- Type .35 into the first row of P.
- Choose **Rows>Add Rows** and specify **# Rows to add: 30**. You must have a table with n+1 rows for the calculations to work properly. A portion of the resulting table is shown below.

EX5-073 using BINOMIAL

5 Cols / 31 Rows	p	x	nCx	Pr (X = x)	Pr(X ≤ x)
1	0.35	0	1	0.000002	0.000002
2	•	1	30	0.000039	0.000042
3	•	2	435	0.000308	0.00035
4	•	3	4060	0.001547	0.001896
5	•	4	27405	0.005621	0.007518
6	•	5	142506	0.01574	0.023258
7	•	6	593775	0.035314	0.058572
8	•	7	2035800	0.065196	0.123768

Notice that some of the values of nCx are rather large since there are so many ways that the middle values of x are possible.

To see the pdf or cdf distribution, you can use **Fit Y by X** plotting x versus either Pr(X = x) or Pr(X ≤ x) as Y.

Exercise 5.76 (pg. 277)

Exercise 5.76 is completed with precisely the same steps as in exercise 5.73, above. The only difference is the value of p and the number of rows used.

Exercise 5.90 (pg. 280)

Exercise 5.90 uses the BINOMIAL template to determine the number of defective shirts in a box of n=12 and p=.1 (where p is the probability of an imperfect shirt). Alternately, you could do p=.9 where p would then be the probability of a single shirt being perfect. Use the BINOMIAL template documented on page 150 in **Appendix: Distribution Calculators: BINOMIAL**. For an example of its use, see exercises 5.73 and 5.76 above. The resulting data table is shown below.

EX5-90a using BINOMIAL

5 Cols / 13 Rows	p	x	nCx	Pr (X = x)	Pr(X ≤ x)
1	0.1	0	1	0.28243	0.28243
2	•	1	12	0.376573	0.659002
3	•	2	66	0.230128	0.88913
4	•	3	220	0.085233	0.974363
5	•	4	495	0.021308	0.995671
6	•	5	792	0.003788	0.999459
7	•	6	924	0.000491	0.99995
8	•	7	792	0.000047	0.999997
9	•	8	495	0.000003	1
10	•	9	220	1.604e-7	1
11	•	10	66	5.346e-9	1
12	•	11	12	1.08e-10	1
13	•	12	1	1e-12	1

For parts b, c, e and f, read off the appropriate values for Pr(X = x) or Pr(X ≤ x).

Part h simulates 200 trials of randomly sampling from a binomial distribution with the same n and p as specified in part a. Use the RANBIN template documented in **Appendix: RANBIN** on page 144. The first few simulated samples from one possible set of values is shown below.

EX5-90h using RANBIN

4 Cols / 200 Rows	# samples per trial	Trial	Sample	Binomial (n,p)
1	200	1	1	1
2	•	1	2	2
3	•	1	3	0
4	•	1	4	2
5	•	1	5	1

The mean, standard deviation, and histogram are parts of the default display of the **Distribution of Y** platform. The default display is on the left, below.

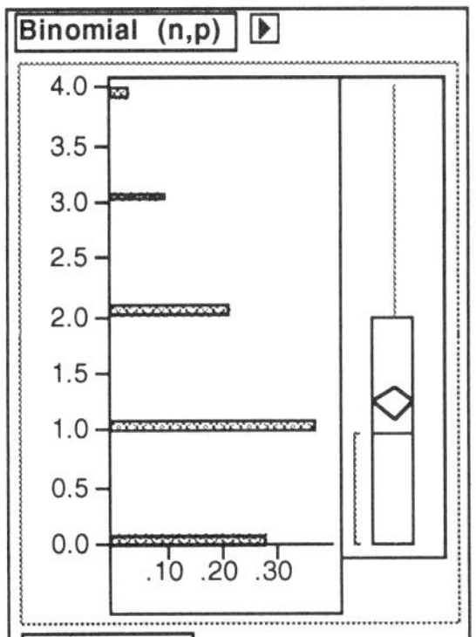

Moments

Mean	1.2550
Std Dev	1.0936
Std Error Mean	0.0773
Upper 95% Mean	1.4075
Lower 95% Mean	1.1025
N	200.0000
Sum Weights	200.0000

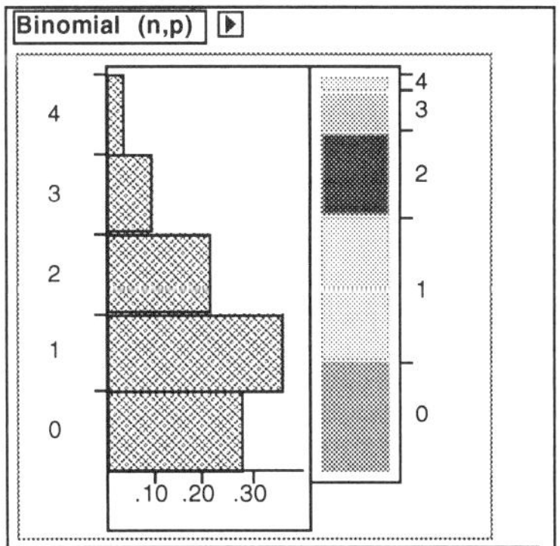

Frequencies

Level	Count	Probability	Cum Prob
0	56	0.28000	0.28000
1	73	0.36500	0.64500
2	43	0.21500	0.86000
3	20	0.10000	0.96000
4	8	0.04000	1.00000
Total	200		
	5 Levels		

The **Distribution of Y** result with the variable marked as nominal is shown on the right. This is a good example of a situation where you may want to view both displays. If you are interested in the average number of imperfect shirts in a box, the continuous display is useful. If you are interested in the number of boxes with zero defects, 1 defect, 2 defects, and so on, the nominal display is the one you need. See the steps documented in **Chapter 2: Using Distribution of Y to describe a variable** on page 22 to add the probability axis to the histogram.

Chapter 6
Normal Probability Distributions

Calculating area under the curve

Chaptcr 6 discusscs continuous distributions and focuscs on thc most commonly used distribution, the standard normal distribution. The area under the curve of the normal probability distribution, the pdf, is called the cumulative normal distribution, the cdf.

Note: Table 3 in ES7 calculates a different area under the curve.

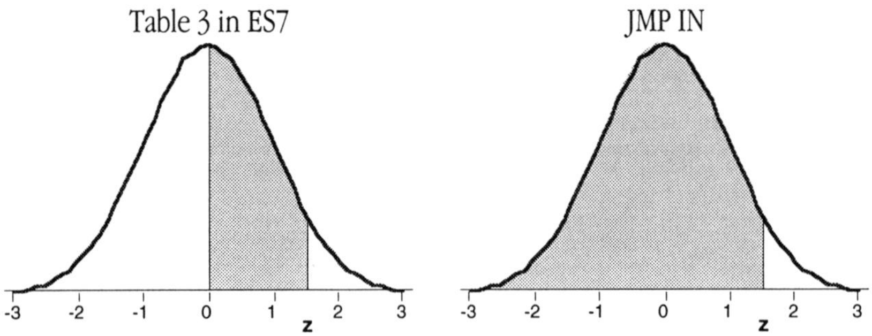

In this chapter you explore what it means for something to be normally distributed and how you can tell if, in a particular case, it's true.

Specific Applications in ES7

Generating a normal distribution (pg. 305)

Exercise 2.137, shown on page 42 in this text, gives detailed instructions on the use of the RANNOR template for generating random samples from a specified normal distribution. Also see the instructions in **Appendix: RANNOR** on page 142. One set of values is shown below for 12 trials each with 100 samples.

EX6-038 using RANNOR

4 Cols / 1200 Rows	# samples per trial	Trial	Sample	Normal (50, 12)
1	100	1	1	33.26013
2	•	1	2	62.37927
3	•	1	3	21.50227
4	•	1	4	48.10904
5	•	1	5	57.53491
6	•	1	6	27.98586
7	•	1	7	64.96886
8	•	1	8	58.80313
9	•	1	9	57.10605
10	•	1	10	56.14428

Calculating values from the normal distribution (pg. 305)

If you have just a few values from a specific normal distribution, it is easiest to use the PROBNORM JMP template documented in **Appendix: Distribution Calculators: PROBNORM** on page 151. To calculate the large numbers required in exercise 6.39 it is easiest to use Copy and Paste to add a new calculated column to the data table of values. The steps to add this calculated column follow.

Open the PROBNORM template and put the pdf formula on the clipboard.

- Choose **File>Open** and specify the file PROBNORM.
- Double-click on the column pdf.
- Click on the mini-formula display button in the Column Information window. This brings up the JMP IN calculator.
- Select the entire formula by clicking on the large vertical bar.
- Choose **Edit>Copy**. Close the calculator window.

Now add the formula into the data table built in exercise and modify the formula so that it calculates the pdf for the correct normal distribution.

- Make the data table (EX6-038 in this case) the front window by clicking on it or by choosing it from the Window menu.
- Choose **Cols>New Col**.
- Specify **Col Name: pdf**, and specify **Source: Formula**.
- When you press **OK** the JMP IN calculator appears. Choose **Edit>Paste**.

- A dialog results, asking for the name of the observed value in this data table. Choose the correct variable (**Normal (50,12)** in this exercise).
- Modify the values of the formula. Change ***Mean*** and ***SD***. The final formula for completing exercise 6.39 appears below.

$$\begin{array}{|l} \boldsymbol{Mean} \Leftarrow 50 \\ \boldsymbol{SD} \Leftarrow 12 \\ \text{results normDensity}\left(\frac{(Normal\ (50,\ 12)-\boldsymbol{Mean})}{\boldsymbol{SD}}\right) \end{array}$$

- Close the calculator window. The normal probabilities calculate.

EX6-039 using the PROBNORM formula

5 Cols / 1200 Rows	per trial	Trial	Sample	Normal (50, 12)	pdf
1	100	1	1	33.26013	0.15078
2	•	1	2	62.37927	0.234...
3	•	1	3	21.50227	0.023...
4	•	1	4	48.10904	0.39402
5	•	1	5	57.53491	0.327...
6	•	1	6	27.98586	0.074...
7	•	1	7	64.96886	0.183...
8	•	1	8	58.80313	0.304...
9	•	1	9	57.10605	0.334...
10	•	1	10	56.14428	0.34993

Plotting normal probabilities (pg. 305)

Exercise 6.40 plots the data from exercise 6.39. Simply use the **Graph>Overlay Plots** platform with Normal (50, 12) as X and pdf as Y.

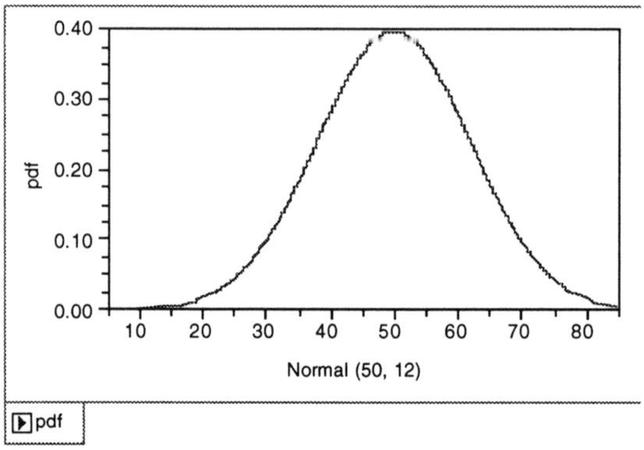

The overlay plot automatically connects the points. To display only the curve, uncheck the ▶ **Show Points** option in the popup menu below the graph.

Exercise 6.57 (pg. 308)

To calculate cumulative probabilities from a specified normal distribution, use the NORMCDF template documented in Appendix: NORMCDF on page 152. The data table to answer part a is shown below.

EX6-057 using NORMCDF

7 Cols / 1 Rows	Mean	Std Dev	X	z	Pr(Z≤z)	Pr(Z>z)	pdf
1	584	37.3	525	-1.58	0.0569	0.9431	0.1142

The value in Pr(Z ≤ z) is cumulative probability.

Exercise 6.58 (pg. 308)

As in exercise 6.57, use the **NORMCDF** template to calculate a range of z values and associated normal probability. The simplest way to do this in this case is to type in the **x** values from 0.0 to 0.5 in intervals of 0.1. Use **Mean**=0 and **Std Dev**=1 and the **z** scores calculate properly.

EX6-058 using NORMCDF

8 Cols / 6 Rows	Mean	Std Dev	X	z	Pr(Z≤z)	Pr(Z>z)	pdf	Table 3 ES7
1	0	1	0	0.00	0.5000	0.5000	0.3989	0.0000
2	•	•	0.1	0.10	0.5398	0.4602	0.3970	0.0398
3	•	•	0.2	0.20	0.5793	0.4207	0.3910	0.0793
4	•	•	0.3	0.30	0.6179	0.3821	0.3814	0.1179
5	•	•	0.4	0.40	0.6554	0.3446	0.3683	0.1554
6	•	•	0.5	0.50	0.6915	0.3085	0.3521	0.1915

To convert the JMP IN normal cdf values to those of ES7 Table 3, use the following formula.

$$Pr(Z \leq z) - 0.5$$

This formula works only for positive z values.

Exercise 6.59 (pg. 308)

In part a, use the **NORMCDF** template and either type in the **x** values from $\mu - 4\sigma$ to $\mu + 4\sigma$ in half standard deviation units or use the JMP calculator. The JMP data table for parts a, b, and c follow.

EX6-059 using NORMCDF

6 Cols / 17 Rows	Mean	Std Dev	X	z	Norm CDF	PROBNORM
1	100	16	36	-4.00	0.0000	0.0001
2	•	•	44	-3.50	0.0002	0.0009
3	•	•	52	-3.00	0.0013	0.0044
4	•	•	60	-2.50	0.0062	0.0175
5	•	•	68	-2.00	0.0228	0.0540
6	•	•	76	-1.50	0.0668	0.1295
7	•	•	84	-1.00	0.1587	0.2420
8	•	•	92	-0.50	0.3085	0.3521
9	•	•	100	0.00	0.5000	0.3989
10	•	•	108	0.50	0.6915	0.3521
11	•	•	116	1.00	0.8413	0.2420
12	•	•	124	1.50	0.9332	0.1295
13	•	•	132	2.00	0.9772	0.0540
14	•	•	140	2.50	0.9938	0.0175
15	•	•	148	3.00	0.9987	0.0044
16	•	•	156	3.50	0.9998	0.0009
17	•	•	164	4.00	1.0000	0.0001

The formula for x is:

count(from (100–4•16), to (100+4•16), in n steps, 1 time)

It is easiest to copy the formula for normal pdf from the PROBNORM JMP template. The formula for PROBNORM is:

normDensity(z)

To graph the normal probability distribution curve for N(100, 16) use the Fit Y by X platform. For parts d, e, and f see the instructions for exercise 2.137 on page 42. Use the RANNOR template documented on page 142.

Exercise 6.60 (pg. 308)

The most straightforward way to generate random normal values is to use the RANNOR template documented on page 142. To generate 10 trials, all with 100 sample values, specify # samples per trial=100 and create a data table with 10 • 100=1000 rows. A portion of a representative data table follows.

EX6-060 using RANNOR

4 Cols / 1000 Rows	# samples per trial	Trial	Sample	Normal (200, 25)
1	100	1	1	245.1206
2	•	1	2	198.0021
3	•	1	3	209.9144
4	•	1	4	172.9171
5	•	1	5	255.9574

In the above table, the last column has been renamed to **Normal (200, 25)** and its formula has a specified mean and standard deviation. In order to draw 10 histograms on the same scale and calculate 10 sets of descriptive statistics it is convenient to rearrange the data table so that instead of the 10 sets of 100 values being stacked on top of one another, there are 10 columns each with 100 rows. After the above table has been created, follow these steps.

- Choose **Tables>Split Columns**.
- Specify the column to be split by selecting **Normal (200,25) >Split>**.
- Specify the values to be used to identify the new columns by selecting **Trial >Col ID>**.
- Press **OK**. A new data table with 100 rows results.

EX6-060 after Split Columns

12 Cols / 100 Rows	ample	1	2	3	4	5	
1	1	245.1206	206.3147	152.2443	176.7085	196.9058	15
2	2	198.0021	217.4765	130.4201	242.6345	180.3528	17
3	3	209.9144	198.7129	209.5388	245.7013	198.3646	20
4	4	172.9171	204.8805	144.7405	240.815	209.0062	15
5	5	255.9574	162.2055	222.7315	173.4086	205.091	17
6	6	184.3942	190.8739	251.1269	205.8727	269.0939	23
7	7	212.8414	208.0802	205.0382	213.9283	195.1078	21
8	8	197.8348	175.7696	209.9483	213.7672	180.8071	21

- Choose **Analyze>Distribution of Y**.
- Specify ten Y columns by selecting **>Y> 1**, **2**, **3**, **4**, **5**, **6**, **7**, **8**, **9**, **10** . The histograms do not automatically have the same scaling.
- Choose ☑ **Uniform Axes** from the popup menu at the bottom left border of the window. Two of the subsets are shown below.

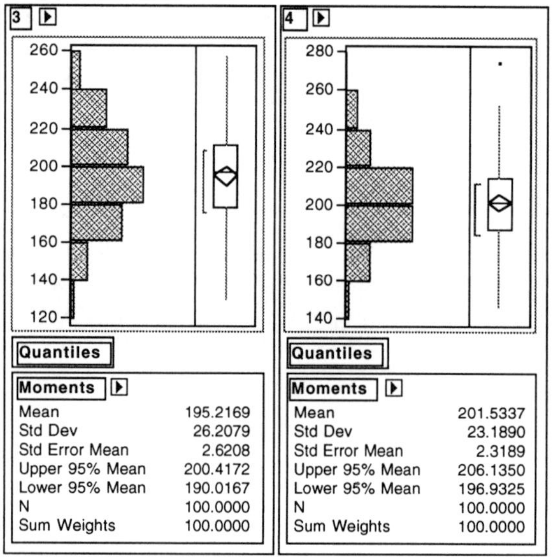

The descriptive statistics are automatically displayed.

Exercise 6.75 (pg. 319)

To calculate the probability distribution for various binomial distributions, use the BINOMIAL template documented on page 150 in **Appendix: Distribution Calculators: BINOMIAL**. For an example of its use, see exercises 5.73, 5.76, or 5.90. The resulting data table for part a. is shown below.

EX6-075 using BINOMIAL

5 Cols / 11 Rows	p	x	nCx	Pr (X = x)	Pr(X ≤ x)
1	0.3	0	1	0.028248	0.028248
2	•	1	10	0.121061	0.149308
3	•	2	45	0.233474	0.382783
4	•	3	120	0.266828	0.649611
5	•	4	210	0.200121	0.849732
6	•	5	252	0.102919	0.952651
7	•	6	210	0.036757	0.989408
8	•	7	120	0.009002	0.99841
9	•	8	45	0.001447	0.999856
10	•	9	10	0.000138	0.999994
11	•	10	1	0.000006	1

To look at the distributions, use the **Graph>Overlay Plot** with x as the X role and Pr(X = x) as Y.

Computational note: For part b and c, *n* is large. Depending upon the speed of your machine, it can take a while to calculate the required probabilities. Be sure to type the new value for P *before* choosing **Rows>Add Rows**.

Exercise 6.107 (pg. 322)

Use the NORMCDF template for this exercise. You are given the Std Dev (0.2), and X setting (6.0) and asked to find the Mean value that results in the Norm CDF ≥ 0.99.

NORMCDF

5 Cols / 1 Rows	Mean	Std Dev	X	z	Norm CDF
1	6	0.2	6	0.00	0.5

By typing various values into Mean, you'll be able to come close.

Chapter 7
Sample Variability

Sampling from a known distribution

The first step in understanding how to deduce information about a population based on a sample is observing patterns of typical variability. Simple cases are bridges to the more real-world situations covered in later chapters. Again, one way to see the kind of things that can happen in sampling is to observe the behavior of computer simulations. Fortunately, in a wide array of cases, the central limit theorem says that sample means are well behaved. Understanding how a sampling distribution behaves allows you to make predictions about the corresponding population parameter. See JSS chapter 5 for information on univariate distributions: one variable, one sample.

Specific Applications in ES7

Exercise 7.11 (pg. 333)

To generate random integers, use the RANINT template documented in **Appendix: Random Number Generation Templates: RANINT** on page 146. Type in # samples per trial = 5. Specify the minimum value (0) and the maximum value (9) in the formula for Digits (L,H). Generate a table with 500 rows.

EX7-011 using RANINT

4 Cols / 500 Rows	# samples per trial	Trial	Sample	digits [L,H]
1	5	1	1	7
2	•	1	2	7
3	•	1	3	4
4	•	1	4	9
5	•	1	5	0
6	•	2	1	1
7	•	2	2	4

JMP IN can calculate summary statistics separately for each trial. After the above table is generated, do the following.

Choose **Tables>Group Summary**.

- Highlight **Trial** in the scrolling list of variable on the left of the dialog.
- Select **>Group>** and the grouping variable appears on the top-right.
- Highlight **digits [L, H]** in the scrolling list of variable on the left of the dialog.
- Select **>Stats>▼Mean** and an entry should appear in the scrolling list on the right.
- Click **OK** and the group/summary table appears. A portion appears below.

RANINT by Trial

3 Cols / By-Mode Off / 100 Rows	Trial	N Rows	Mean(digits [L,H])
1	1	5	5.4
2	2	5	3.8
3	3	5	3.8
4	4	5	3.6
5	5	5	3.8
6	6	5	4.6
7	7	5	5.2
8	8	5	4.4

- Choose **Analyze>Distribution of Y** and the histogram and other summary statistical information on the sample means appears. In **Chapter 2: Specific Applications in ES7: Normal probability paper and testing for normality** on page 41 there is additional information on graphically assessing normality.

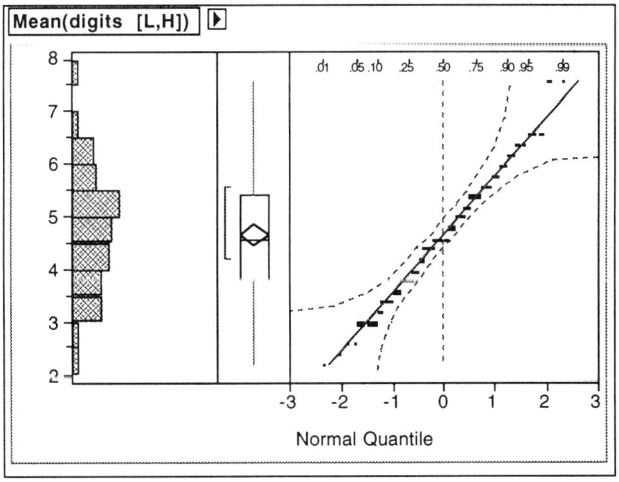

Exercise 7.12 (pg. 334)

The steps for doing exercise 7.12 are the same as the steps for exercise 7.11. The only difference is that the **RANNOR** template is used instead of RANINT.

Exercise 7.19 (pg. 341)

The steps for doing exercise 7.19 are the same as the steps for exercise 7.11 and 7.12. Use the **RANNOR** template.

Exercise 7.33 (pg. 346)

The **NORMCDF** template calculates probability values for the normal distribution. See page 152 for documentation of its use.

NORMCDF

7 Cols / 3 Rows	Mean	Std Dev	X	z	Pr(Z≤z)	Pr(Z>z)	pdf
1	5	1	4	-1.00	0.1587	0.8413	0.2420
2	•	•	5	0.00	0.5000	0.5000	0.3989
3	•	•	6	1.00	0.8413	0.1587	0.2420

Note the value of **Std Dev** used. Since $\Pr(Z \leq -1) = \Pr(X < 4)$ and X, in this case is $\bar{\mathbf{X}}$, $\Pr(\bar{\mathbf{X}} < 4) = 0.1587$. Also, $\Pr(\bar{\mathbf{X}} < 6) = 0.8413$. So, $\Pr(4 < \bar{\mathbf{X}} < 6) = 0.8413 - 0.1587 = 0.6826$. Or, more simply, recall from the empirical rule of thumb for the normal distribution in ES7 pgs. 104-107 that approximately 68% of the data are within one standard deviation.

Use the **RANNOR** template to generate 100 trials of sample size 4 for a normal distribution with mean 5 and standard deviation 2. Then use **Tables>Group** Summary to calculate the means of each of the 100 trials. A portion of an example summary table appears below.

EX7-033 by Trial

3 Cols / By-Mode Off / 100 Rows	Trial	N Rows	Mean(Normal (5, 2))
1	1	4	3.34573
2	2	4	2.546202
3	3	4	4.329094
4	4	4	6.304846
5	5	4	4.914811
6	6	4	5.686958

Rows in the data table are highlighted by shift-clicking on bars in the **Distribution of Y** histogram. Highlighting the values in the two bars corresponds to selecting all of the means with values between 4 and 6.

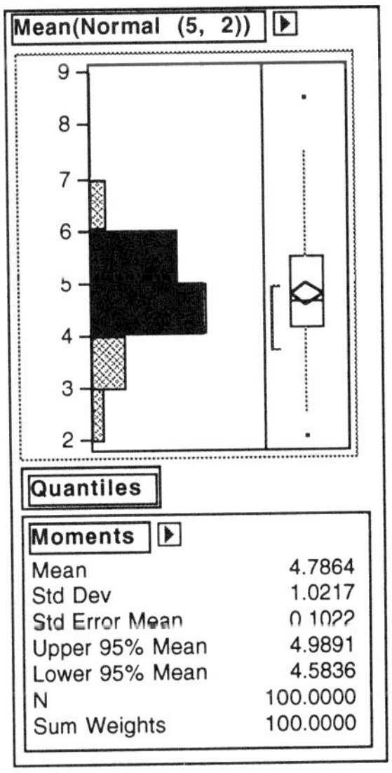

One way to count the number of trials with values between 4 and 6 is to use **Tables>Subset** to extract only those rows with highlighted values.

Subset of EX7-33 by Trial

3 Cols / 74 Rows	Trial	N Rows	Mean(Normal (5, 2))
1	3	4	4.329094
2	5	4	4.914811
3	6	4	5.686958
4	7	4	4.065538
5	8	4	5.780068

The above simulation had 74/100 rows within the required range. Everyone will get different results.

Exercise 7.50 (pg. 349)

The main difference between this exercise and the other JMP IN exercises in this chapter is that in addition to calculating a mean, the sum is used.

- Select **>Stats>▼Sum** to calculate totals for each trial.

Exercise 7.51 (pg. 350)

Use the BINOMIAL template to construct the probability distribution needed. Use RANBIN to generate 200 trials of sample size 25 from binomial (16, 0.5).

EX7-051 using RANBIN

4 Cols / 5000 Rows	# samples per trial	Trial	Sample	Binomial (16,0.5)
1	25	1	1	5
2	•	1	2	4
3	•	1	3	10
4	•	1	4	11
5	•	1	5	9
6	•	1	6	6
7	•	1	7	6
8	•	1	8	11
9	•	1	9	11

Use the steps in exercise 7.11 to calculate summary statistics for the groups.

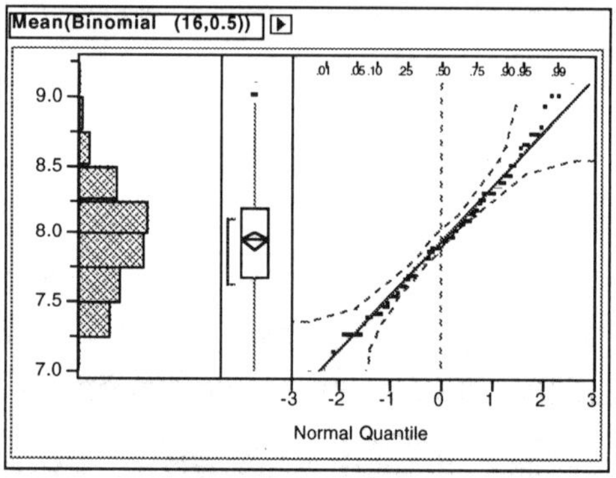

Exercise 7.52 (pg. 350)

The steps for this exercise are identical to those for exercise 7.51. The only difference is that a different n and p are used to specify the binomial distribution.

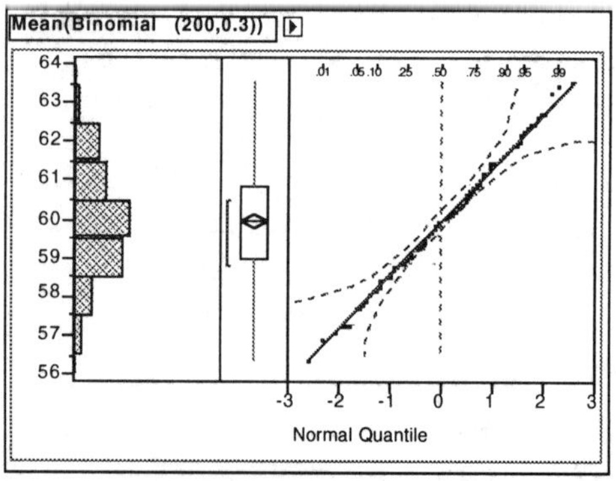

Chapter 8
Introduction to Statistical Inferences

Estimates, Confidence Intervals, and Hypothesis Testing

Estimating population characteristics based on sampling results and making decisions about the population based on sampling information is the core application of statistics in the real world. How good is an estimate? How sure are you that you've made the correct decision? We begin by considering the situation where we know about the amount of variability in the population.

Two cases are covered in the Specific Applications in this chapter. How do you form confidence intervals and do hypothesis testing with data values? And, how do you do it when all you have is summary statistics such as a sample mean? Also see JSS chapter 5 for information on univariate distributions: one variable, one sample.

Specific Applications in ES7

Finding the confidence interval for mean (pg. 373)

When σ is known and other assumptions hold, formula 8-1 gives confidence interval on the mean. Use the FORM8-1 template, documented in **Appendix: Data Analysis Templates: FORM8-1** on page 134. For instance, the data for Illustration 8-3 and resulting 90% confidence interval are shown below.

ILL8-3 using FORM8-1

7 Cols / 1 Rows	Sigma	n	X bar	alpha (a)	SE Mean	Lower CI	Upper CI
1	2.87	40	3.975	0.1	0.453787	3.228587	4.721413

Completing the hypothesis test for mean (pg. 399)

Case 1: Data values If you have a set of data values the preferred way to test a one-sample hypothesis test on the mean is by using the **Distribution of Y** platform. See **Chapter 2: Using Distribution of Y to describe a variable: Optional displays and reports**, which begins on page 26 for an initial discussion of **Common ▶ popup**

options. The **Test mean=value** option performs hypothesis tests using either the z or t distribution, as well as using nonparametric methods. For example, using the 40 integers in Illustration 8-3, do the following.

- Choose **Analyze>Distribution of Y** and specify the Y variable.
- Choose ▶ **Test mean=value** from the popup at the top of the histogram. The **Test Mean** dialog appears.
- Enter a value for **Specify Hypothesized Value for Mean: 4.5**.
- Enter a value for **... true standard deviation...: 2.87**. If you specify σ, JMP IN automatically uses the z statistic.
- Click **OK**. The **Test Mean=value** report appears.

Test Mean=value		
Hypothesized Value		4.5
Actual Estimate		3.975
Using Std Dev of		2.87
	z Test	
Test Statistic	-1.1569	
Prob > |z|	0.2473	
Prob > z	0.8763	
Prob < z	0.1237	

Note that three p-values appear. Which one you use depends on whether you have chosen a two-sided Ha: or a one-sided Ha: and the direction of your one-sided test. Prob >|z| is the two-sided p-value which describes how likely it is to be different in either direction. Prob > z is the one-sided alternative of getting an even greater z value if there were no difference. Prob < z is the one-sided alternative of getting an even smaller z value if there were no difference. In the case above, the decision under all three scenarios would be to "fail to reject Ho:."

Case 2: Summary statistics If you only have the sample mean, sample size, hypothesized mean and known sigma, use the FORM8-3 JMP template. It is documented on page 135. For example, for the data used in Illustration 8-3 here are the results.

ILL8-3 using FORM8-3

8 Cols / 1 Rows	Mu	Sigma	n	X Bar	z star	Prob |z|	Prob > z	Prob < z
1	4.5	2.87	40	3.975	-1.157	0.247301	0.87635	0.12365

Exercise 8.85 (pg. 401) and Exercise 8.86 (pg. 402)

In exercise 8.85 and 8.86 all you have is summary statistics. Use the FORM8-3 template.

EX8-085 using FORM8-3

8 Cols / 1 Rows	Mu	Sigma	n	X Bar	z star	Prob |z|	Prob > z	Prob < z
1	525	60	38	512.14	-1.321	0.186422	0.906789	0.093211

Exercise 8.100 (pg. 415)

In Table 8-10 the sample means from twenty trials are given. The data comes from random digits where each sample is of size 40. In exercise 7.11 a similar set of random values was generated (see page 76).

- Use the RANINT template documented in on page 146. To complete exercise 8.100 enter # samples per trial = 40. Specify the minimum value (0) and the maximum value (9) in the formula for Digits (L,H). Generate a table with 40 rows. A portion of such a table appears below.

EX8-100 using RANINT

4 Cols / 40 Rows	# samples per trial	Trial	Sample	digits [L,H]
1	40	1	1	7
2	•	1	2	3
3	•	1	3	5
4	•	1	4	9
5	•	1	5	9

- Change the modeling type of digits [L,H] to continuous.
- Now all you have to do is use the **Distribution of Y** platform, choose Test mean=value, and fill in the dialog (see **Completing the hypothesis test for mean** beginning in this chapter on page 81).

Test Mean=value		
Hypothesized Value		4.5
Actual Estimate		5.45
Using Std Dev of		2.87
	z Test	
Test Statistic	2.0935	
Prob > |z|	0.0363	
Prob > z	0.0182	
Prob < z	0.9818	

The 40 random digits generated in the above case would end in a decision to "reject Ho."

Exercise 8.115 and Exercise 8.116 (pg. 418)

See Exercise 8.85 and Exercise 8.86, in this chapter.

Exercise 8.151 (pg. 425)

The 25 values in exercise 8.151 are on the student disk. After reading in the data, use Group/Summary to calculate the sample mean. Then simply use the FORM8-1 template to calculate the 95% confidence interval.

- With the EX8-151 data table as the front window, choose **Tables>Group/Summary**.
- Select X **>Stats>▼Mean** and an entry should appear on the right.
- Click **OK** and the group/summary table appears.
- Click on the value in the data table (the mean of X). It appears as below.

1 Cols / 1 Rows	Mean(X)
1	129.0212

- Choose **Edit>Copy** to put the observed mean value on the clipboard.
- Choose **File>Open** and select **FORM8-1**.
- Click on the value in second column in the data table. It appears as below.

EX8-151 using FORM8-1

7 Cols / 1 Rows	n	Mean (X)	Std Dev (X)	alpha (a)	SE Mean	Lower CI	Upper CI
1	•	•	•	0.05	•	•	•

- Choose **Edit>Paste** to put the observed mean value into the table.
- Type the other values given in the exercise for N, Std Dev. The final results:

EX8-151 using FORM8-1

7 Cols / 1 Rows	n	Mean (X)	Std Dev (X)	alpha (a)	SE Mean	Lower CI	Upper CI
1	25	129.0212	10	0.05	2	125.1013	132.9411

Exercise 8.153 (pg. 425)

The data in the exercise are on the student disk. After reading in the data, use the **Distribution of Y** platform, choose Test mean=value, and fill in the dialog.

Test Mean=value

Hypothesized Value	18
Actual Estimate	17.2168
Using Std Dev of	4
	z Test
Test Statistic	-1.0360
Prob > |z|	0.3002
Prob > z	0.8499
Prob < z	0.1501

Chapter 9
Inferences Involving One Population

Sigma Unknown

When working in the real world, you rarely know the amount of variability in the population of interest. This chapter covers one-sample problems where you are interested in the mean of a continuous measurement, an observed sample proportion of a yes/no variable, or the amount of variability of a continuous measurement. Also see JSS chapter 5 for information on univariate distributions: one variable, one sample.

Specific Applications in ES7

Calculating values from Student's t distribution (pg. 437-438)

The JMP IN calculator has probability functions for all the major distributions. To calculate the cdf (p-value) when given a (Student's) t-value and degrees of freedom, use the TCDF template. To calculate the t-value when given the cdf (p-value) and degrees of freedom, use the TDIST template. These are documented in the **Appendix: Distribution Calculators** on page 154.

The example from ES7 pg. 438 for calculating the cdf is shown below.

TCDF

3 Cols / 1 Rows	t	df	Pr(T≤t)
1	-1.74	17	0.049965

To use the TCDF template, simply type in values for t and df and the probabilities automatically calculate.

The example for calculating the inverse cdf (the t value) is shown on the next page.

TDIST

3 Cols / 1 Rows	Pr(T≤t)	df	t
1	0.05	17	-1.73961

To use the TDIST template, simply type in values for Pr(T≤t) and df and the distribution values automatically calculate.

Finding the confidence interval for mean (pg. 440)

When σ is not known, Student's t distribution can be used to form a confidence interval for μ. In JMP IN there are two methods to obtain this interval. If you have data values and you want the 95% confidence interval, the **Distribution of Y** platform makes the problem a trivial one. If all you have is summary statistics (the observed mean, standard deviation, and sample size), then the calculations in formula 9-1 are used.

Case 1: Data values If you have a set of data values, the preferred way to calculate the 95% confidence interval or to test hypotheses about the mean is by using the **Distribution of Y** platform. For instance, exercise 9.28 uses 21 data values saved on the data disk. After reading these values into JMP IN, follow the steps below.

- Choose **Analyze>Distribution of Y**.
- Specify WEIGHT>**Y>**. The default **Moments** report shows the 95% confidence interval about the mean.

Moments ▶	
Mean	11.48571
Std Dev	0.47464
Std Error Mean	0.10358
Upper 95% Mean	11.70177
Lower 95% Mean	11.26966
N	21.00000
Sum Weights	21.00000

The 95% confidence interval is 11.27 - 11.70, rounded to 2 decimals.

If you want to test a hypothesis about the mean of a set of data values, there is just one additional step. In the above data you may want to test the hypothesis that the mean is 12 ounces.

- Choose ▶ **Test mean=value** from the popup at the top of the histogram.

The Test Mean dialog appears.

- Enter a value for **Specify Hypothesized Value for Mean: 12**.
- Click **OK**. The **Test Mean=value** report appears.

Test Mean=value		
Hypothesized Value		12
Actual Estimate		11.4857
	t Test	
Test Statistic	-4.9653	
Prob > |t|	<.0001	
Prob > t	1.0000	
Prob < t	<.0001	

It seems rather unlikely that the average can is filled to 12 ounces.

Case 2: Summary statistics If you only have the sample mean, sample standard deviation, sample size, and the hypothesized mean, use the **FORM9-1** JMP template to calculate a confidence interval. It is documented on page 136. For example, for the data used in Illustration 9-4, here are the results.

ILL9-4 using FORM9-1

7 Cols / 1 Rows	n	Mean (X)	Std Dev (X)	alpha (a)	SE Mean	Lower CI	Upper CI
1	20	6.87	1.76	0.05	0.393548	6.046295	7.693705

Completing the t-test of mean (pg. 444)

Just as with the confidence interval procedure above, if you have data values and you want the 95% confidence interval, the **Distribution of Y** platform makes the problem easy. If all you have is summary statistics (the hypothesized μ, an observed mean, standard deviation, and sample size), then the calculations in formula 9-2 are used.

Case 1: Data values See the instructions for **Finding the confidence interval for mean** on page 86.

Case 2: Summary statistics If you only have an hypothesized mean, the sample mean, sample standard deviation, sample size, and the hypothesized mean, use the **FORM9-2** JMP template to calculate a confidence interval. It is documented on page 136. For example, for the data used in Illustration 9-5 here are the results.

ILL9-5 using FORM9-2

8 Cols / 1 Rows	Mu	n	X Bar	Std Dev	t star	Prob |t|	Prob > t	Prob < t
1	4.9	22	5.1	1.17	0.802	0.43166	0.21583	0.78417

Exercise 9.27 (pg. 449)

To calculate a 90% confidence interval from summary statistics, use the method described in **Finding the confidence interval for mean** on page 86.

EX9-027 using FORM9-1

7 Cols / 1 Rows	n	Mean (X)	Std Dev (X)	alpha (a)	SE Mean	Lower CI	Upper CI
1	13	75.69231	14.53951	0.1	4.032535	68.50517	82.87945

Exercise 9.39 (pg. 452)

To test a hypothesis that the mean of the values given in exercise 9.39 is equal to 160, see the instructions for **Finding the confidence interval for mean: Case 1: Data values** in the previous section. The data is on the student disk in EX9-039. The results appear in the following report.

Test Mean=value		
Hypothesized Value		160
Actual Estimate		160.25
	t Test	
Test Statistic	0.0541	
Prob > |t|	0.9576	
Prob > t	0.4788	
Prob < t	0.5212	

Exercise 9.42 (pg. 453)

This exercise simulates 10 trials of sample size 100 from normal, rectangular, and exponential distribution. To simulate the required distributions, use the RANNOR, RANUNI, and RANEXP JMP templates documented in the Appendix. The steps for doing each of the three cases are the same. The last case of the skewed population is given below.

- Use the RANEXP template to generate # samples per trial = 10, exponentially distributed random values. Since you want 100 trials, the resulting table has 1000 rows.

EX9-042c using RANEXP

4 Cols / 1000 Rows	# samples per trial	Trial	Sample	Exp (100)
1	10	1	1	12.70
2	•	1	2	14.24
3	•	1	3	3.55
4	•	1	4	192.11
5	•	1	5	22.91
6	•	1	6	2.30
7	•	1	7	59.36

The remaining steps are similar to those for exercise 7.11 shown on page 76.

- Choose **Tables>Group Summary**.
- Select Trial **>Group>** and the grouping variable appears on the right.
- Select Exponential (100) **>Stats>▼Mean** and an entry should appear in the scrolling list on the right.
- Select Exponential (100) **>Stats>▼Std Dev** and another entry should appear in the scrolling list on the right.
- Click **OK** and the group/summary table appears.
- Add a new column called t star with the following formula.

$$\frac{Mean(Exp\ (100))-100}{\left(\frac{Std\ Dev(Exp\ (100))}{\sqrt{N\ Rows}}\right)}$$

- Remove the [Y] role for STD DEV (EXP 100) and Specify a [Y] role for T STAR.
- A portion of the final data table appears below.

RANEXP by Trial

5 Cols / By-Mode Off / 100 Rows	Trial	N Rows	Mean(Exp (100))	Std Dev(Exp (100))	t star
1	1	10	36.30939	57.22767	-3.51941
2	2	10	61.79044	44.8257	-2.69553
3	3	10	83.54673	83.04521	-0.62652
4	4	10	87.22632	63.49298	-0.63619
5	5	10	88.88789	92.1682	-0.38125
6	6	10	56.18382	50.56105	-2.74043
7	7	10	92.80884	73.44406	-0.30963
8	8	10	72.27244	54.90188	-1.59707

- Choose **Analyze>Distribution of Y** and the histogram and other summary statistical information on the sample means appears. In **Chapter 2: Specific Applications in ES7: Normal probability paper and testing for normality** on page 41.

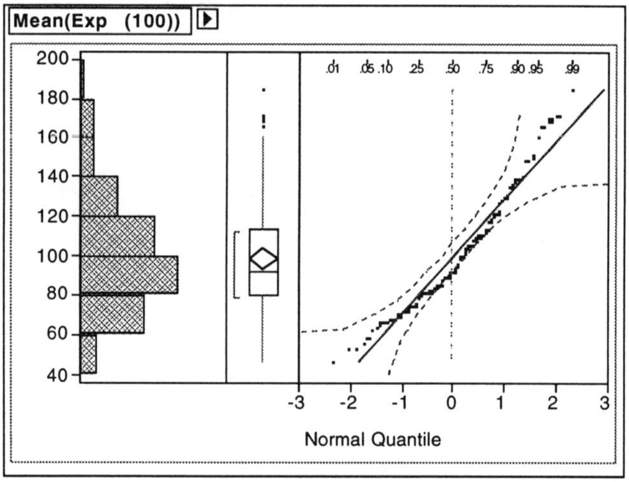

In the distribution above, there does seem to be more large values than would be present in a normal distribution. Your specific results will be different.

Confidence interval for p (pg. 455-456)

To form a confidence interval for the binomial parameter p, given a sample of size n and observed number of success x, use the JMP template FORM9-6 documented on page 137 of the Appendix. For example, the results for Illustration 9-10 are shown below.

FORM9-6

7 Cols / 1 Rows	n	x	alpha (a)	Obs:p'	Std Err	Lower CI	Upper CI
1	200	17	0.1	0.085	0.020	0.053	0.117

The 90% confidence interval for p is 0.053 to 0.117. The reason that the confidence interval in ES7 is slightly different is that JMP IN uses more numerical precision to calculate the values of the observed proportion.

Hypothesis testing about p (pg. 459-461)

To test an hypothesis about the binomial parameter p, ES7 describe a procedure beginning on pg. 459. The resulting formula uses the hypothesized proportion p, the sample size n, and x, the observed number of successes out of n samples. To perform these calculations in JMP IN use the **FORM9-9** template documented on page 138. For instance, the data given in Illustration 9-13 results in the following calculated z★ and p-values.

ILL9-13 using FORM9-9

9 Cols / 1 Rows	Ho:p	n	x	Obs:p'	z star	Prob |z|	Prob > z	Prob < z
1	0.61	350	235	0.6714	2.356	0.018464	0.009232	0.990768

You may notice that the **z star** value calculated above is slightly different than the z★ given on page 460 of ES7. The JMP IN formula for the observed proportion use more decimal places of precision so all of the other calculated values in JMP IN are also more accurate.

If you want to use Method 3 on page 461 of ES7, use the **NORMCDF** template. See the instructions for exercise 6.57 on page 71 on how to do this. The data table to answer part a is shown below.

ILL9-13 using NORMCDF

5 Cols / 1 Rows	Mean	Std Dev	X	z	Norm CDF
1	0	1	2.356	2.36	0.990764

Exercise 9.67 (pg. 467)

Exercise 9.67 looks at 24,000 coin tosses. To calculate the values needed for parts a, b, and c, use the section on **Confidence interval for p** on page 90 of this chapter. For Pearson's experiment the results appear below.

EX9-067 part a using FORM9-6

7 Cols / 1 Rows	n	x	Obs:p'	alpha (a)	Std Err	Lower CI	Upper CI
1	24000	12012	0.5005	0.05	0.003	0.494	0.507

To simulate 24,000 coin tosses, use the **RANCOIN** template documented on page 144 of the Appendix. A portion of one simulation appears below.

EX9-067 part d using RANCOIN

4 Cols / 24000 Rows	C ☐ # samples per trial	N ☐ Trial	N ☐ Sample	C ☐ Coin 0/1
1	24000	1	1	0
2	•	1	2	1
3	•	1	3	0
4	•	1	4	0
5	•	1	5	1

To calculate the number of rows that resulted in a value of 1, use the following steps

- Choose **Tables>Group/Summary**.
- Select Coin 0/1 **>Stats>▼Sum** and the group/summary table appears.

1 Cols / 1 Rows	C Y Sum(Coin 0/1)
1	12071

Whatever sum you get, just type the value into the FORM9-6 template under x and the confidence intervals recalculates.

Exercise 9.68 (pg. 468)

The random binomial values can be simulated using the method described for exercise 7.51 on page 79. Exercise 9.68 simulates a number of combinations of n and p. An example data table for part a is shown below.

EX9-068 using RANBIN

6 Cols / 1000 Rows	C ☐ # samples per trial	N ☐ Trial	N ☐ Sample	C ☐ Binomial (15,0.1)	C Y Obs:p	C Y z star
1	1000	1	1	0	0.00	-1.29
2	•	1	2	4	0.27	2.15
3	•	1	3	1	0.07	-0.43
4	•	1	4	1	0.07	-0.43

The formula for the calculation of the observed proportion, Obs: p, is as follows.

$$\frac{Binomial\ (15,0.1)}{15}$$

Note that a different denominator is used for parts b-e. The z star formula is given below.

$$\left| \begin{array}{l} N \Leftarrow 15 \\ p \Leftarrow 0.1 \\ \text{results } \dfrac{Obs{:}p - \boldsymbol{p}}{\sqrt{\dfrac{\boldsymbol{p}\cdot(1-\boldsymbol{p})}{N}}} \end{array} \right.$$

It also has variables that must be changed for the other parts of the exercise.

Using Chi-square cdf values (pgs. 475, 476, 478, 479)

The area under the curve for the chi-square distribution is calculated using the CHISQCDF template documented in the Appendix on page 155. For instance, the calculation of the p-value for Illustration 9-17 is shown below.

ILL9-17 using CHISQCDF

5 Cols / 1 Rows	df	X	Pr(X≤x)	Pr(X>x)	pdf
1	27	47.25	0.9907	0.0093	0.0024

The p-value is thus 1-.9907 = .0093.

Hypothesis testing for a single variance (pg. 480)

To test an hypothesis on a variance or standard deviation, use the FORM9-10 template documented on page 139 of the Appendix. The example shown in Illustration 9-20 is given below.

ILL9-20 using FORM9-10

6 Cols / 1 Rows	Ho:Sigma	n	Obs:s	chi^2 star	Prob > Chi^2	Prob ≤ Chi^2
1	10	12	14	21.560	0.028012	0.971988

Note that both the left and right areas under the chi-square distribution curve are given in the template.

Exercise 9.111 (pg. 484)

Random values from a chi-square distribution can be simulated using the RANCHISQ template documented on page 149 of the Appendix. For example, a portion of one set of values for df=1 is as below.

EX9-111 using RANCHISQ

4 Cols / 1000 Rows	# samples per trial	Trial	Sample	Chi-square(1)
1	1000	1	1	0.392878
2	•	1	2	0.72812
3	•	1	3	0.810664
4	•	1	4	2.263185
5	•	1	5	0.760053
6	•	1	6	0.130027
7	•	1	7	1.717214

The **Distribution of Y** platform provides all the necessary descriptive information.

Chi-square(1)

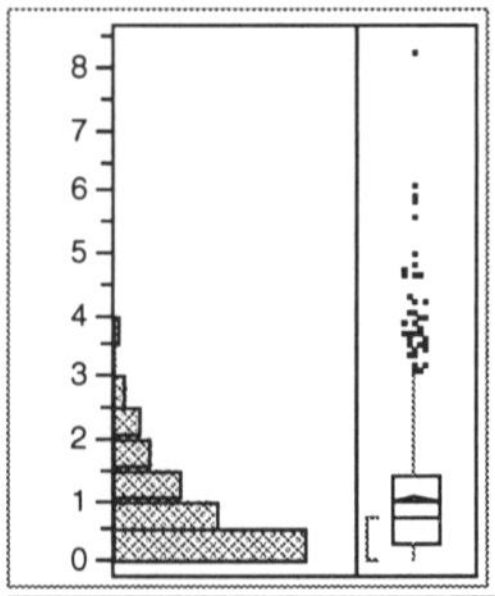

Quantiles

maximum	100.0%	8.2629
	99.5%	5.5958
	97.5%	3.7283
	90.0%	2.3184
quartile	75.0%	1.4149
median	50.0%	0.7100
quartile	25.0%	0.2809
	10.0%	0.1029
	2.5%	0.0304
	0.5%	0.0053
minimum	0.0%	0.0004

Moments

Mean	1.008
Std Dev	1.005
Std Error Mean	0.032
Upper 95% Mean	1.070
Lower 95% Mean	0.946
N	1000.000
Sum Weights	1000.000

Exercise 9.112 (pg. 484)

In exercise 9.112 normal and exponential populations are simulated. This, and a uniform distribution, is precisely what was simulated in exercise 9.42 on pg. 453 of ES7. For instructions on how to simulate samples from these populations and calculate sample statistics, refer to page 88. The exponential simulation is described here. In the simulations for exercise 9.42 the following summary statistics were calculated from Tables>Group/ Summary: N ROWS, Mean(Exp (100)), and Std Dev(Exp 100)). Use these calculated values and the FORM9-10 template to calculate the chi-square statistic for each trial.

- Copy the sample size and sample standard deviation from the Group/Summary table. Highlight the two columns using Option-click (Macintosh) or Alt-click (Windows). Choose **Edit>Copy**.

EX9-112 Using RANEXP by Trial

5 Cols By-Mode Off 100 Rows	Trial	N Rows	N(Exp (100))	Mean(Exp (100))	Std Dev(Exp (100))
1	1	10	10	104.547	132.4666
2	2	10	10	100.562	85.28595
3	3	10	10	109.5776	110.0346
4	4	10	10	148.9446	146.337
5	5	10	10	89.02123	110.0565
6	6	10	10	122.0451	106.7199
7	7	10	10	91.87106	107.605
8	8	10	10	109.2141	102.8264
9	9	10	10	117.8043	100.0083

- Open the FORM9-10 template and delete any rows that it may contain. Choose **Rows>Select>All Rows**. Choose **Rows>Delete Rows**.
- Select the columns n and Obs:s using Shift-click.

EX9-112 using FORM9-10

6 Cols 0 Rows	Ho:Sigma	n	Obs:s	chi^2 star	Prob >

- Choose **Edit>Paste** to add 100 rows of summary values to the empty table.
- Type value for the hypothesized standard deviation into the first row.

EX9-112 using FORM9-10

6 Cols 100 Rows	Ho:Sigma	n	Obs:s	chi^2 star	Prob
1	100	10	132.4666	15.793	
2	•	10	85.28595	6.546	
3	•	10	110.0346	10.897	
4	•	10	146.337	19.273	

- Choose **Analyze>Distribution of Y** and specify Obs:s and chi^2 star as Y variables.

- Use the **RANCHISQ** template to simulate the distribution of a chi-square with df=9 and compare its distribution of that of chi^2 above.

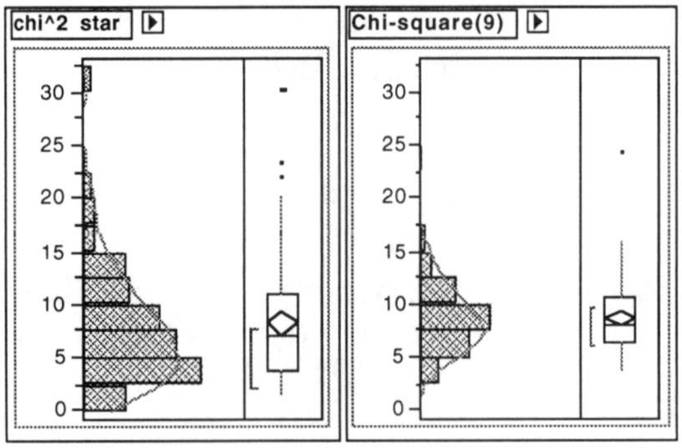

In the example of a normal distribution (not shown here) the chi-square for each sample standard deviation does seem to be distributed the same as a chi-square with df=9.

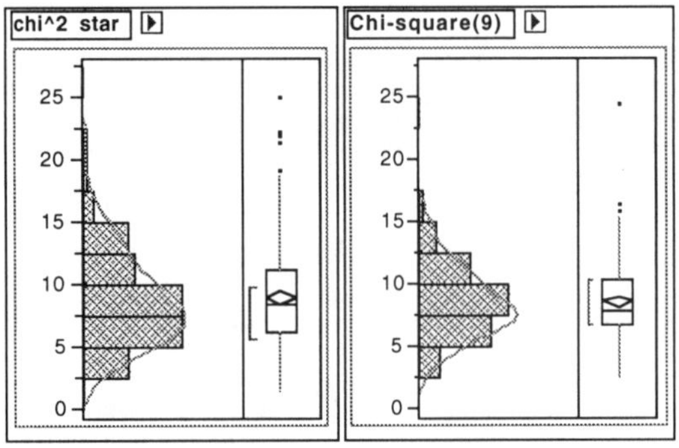

Chapter 10
Inferences Involving Two Populations

Comparing Results from Two Samples

Whether samples are independent or dependent makes a great deal of difference in how you compare groups of data. If the samples are paired (dependent), the Specific Applications comparing dependent means is appropriate. If samples are from two different groups (independent), the Application on inference for two independent means is applicable. The methods for comparing the variability in two groups is covered in chapter 10. Initial methods for comparing proportions are developed in the last section of chapter 10 in ES7, which is a good lead into the more complete methods of chapter 11. Also see chapter 6 in JSS for methods for looking at differences between two means, whether they are from independent or dependent groups.

Specific Applications in ES7

Find confidence interval for difference between two dependent means (pg. 503)

Except for one calculation that must be performed first, there is no difference between calculating confidence intervals here and calculating confidence intervals on the mean, as described in Chapter 9 on page 86. The additional calculation simply involves calculating the difference between the two columns of dependent values (paired data) you wish to contrast. See exercise 10.20, below, for an example.

Hypothesis test for difference between two dependent means (pg. 505) and exercise 10.20 (pg. 507)

The data for exercise 10.20 is on the data disk in EX10-020. However, when you inspect the data file you might see that there is an additional row that must be deleted. After reading in the data and repairing the data table. Follow these steps to calculate a 95% confidence interval.

- Add a new column with a formula which calculates the difference between the two data columns. Choose **Cols>Add Col** and specify **Col name: Diff** (or whatever you want) and **Data Source: Formula**. Specify the following formula.

  AFTER–BEFORE

  Of course, you could specify ***BEFORE-AFTER***. Just keep straight in your mind what a positive difference and a negative difference mean in either case.

- Now, since you have data values and want a 95% confidence interval, follow the instructions in **Chapter 9: Finding the confidence interval for mean: Case 1: Data values** page 86.

Moments ▶	
Mean	6.10000
Std Dev	4.79467
Std Error Mean	1.51621
Upper 95% Mean	9.52993
Lower 95% Mean	2.67007
N	10.00000
Sum Weights	10.00000

Exercise 10.29 (pg. 510)

As in the previous exercise, after the difference is calculated, the method for a hypothesis test on a difference is the same as the method for the hypothesis test on the mean given in **Chapter 9: Completing the t-test of mean: Case 1: Data values** given on page 87. The method simply involves choosing the ▶ **Test mean=value** from the popup at the top of the histogram in the **Distribution of Y** platform. In exercise 10.29 the report appears as below.

Test Mean=value		
Hypothesized Value		0
Actual Estimate		3.2
	t Test	
Test Statistic	3.6923	
Prob > |t|	0.0050	
Prob > t	0.0025	
Prob < t	0.9975	

Inference for two independent means (pg. 522)

The method shown here corresponds to comparing two population means using what ES7 calls "Case 1" on the bottom of pg. 513. Case 1 is the method used by computer software, although there are variations even between different software packages.

Exercise 10.43 (pg. 523)

Data Values: The data on the student disk for exercise 10.43 is oriented so that each independent group's values are stored in separate columns.

EX10-043

2 Cols / 16 Rows	MALES	FEMALES
1	76	76
2	76	70
3	74	82
4	70	90
5	80	68
6	68	60
7	90	62
8	70	68
9	90	80
10	72	74
11	76	60
12	80	62
13	68	72
14	72	•
15	96	•
16	80	•

Although common, this orientation is not suited for an analysis of independent samples in JMP IN. Notice that there is no "pairing" of the values in a row as in the case for the methods for dependent samples. The way the data table should be organized for JMP IN to perform the calculations required is as below.

EX10-43Stacked

2 Cols / 32 Rows	Diastolic	Gender
1	76	MALES
2	76	FEMALES
3	76	MALES
4	70	FEMALES
5	74	MALES
6	82	FEMALES
7	70	MALES
0	00	FEMALES
9	80	MALES
10	68	FEMALES
11	68	MALES

Notice that the values for the MALES and FEMALES columns in the original data table are stacked on top of one another. There were two characteristics of individuals in the study described in exercise 10.43: their gender and their diastolic blood pressure. These two characteristics appear in separate columns and each row corresponds to an individual. The groups to be compared are identified by their nominal values in the GENDER column. The continuous quantitative value that is of interest is in the DIASTOLIC column. In the original data table there were 16 rows and in the new data table there are twice as many rows. Since there are fewer females than males, the last few DIASTOLIC values in the new data table are missing for the last three female rows. To modify the original data table and generate a new data table in the appropriate layout, do the following after making the EX10-043 window the front window.

- Choose **Tables>Stack Columns** and a dialog appears.
- Specify the columns to stack by:
  - Specify **MALES >Stack>** and
  - Specify **FEMALES >Stack>**.
- Name the new **Output table: EX10-43 Stacked**.
- Specify the **Name of Stacked Cols: Diastolic**.
- Specify the **Name of ID Column: Gender**. When the dialog is complete it appears as follows.

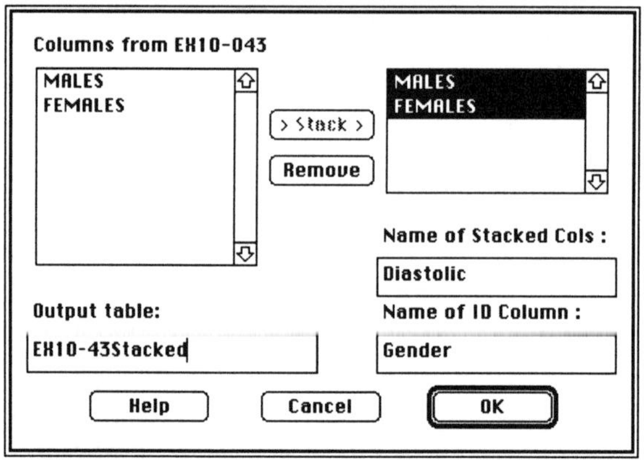

- After completing the dialog, complete the analysis by choosing **Analyze>Fit Y by X**.
- If you have not already done so, specify **Diastolic>Y>** as the response column.
- And, specify **Gender>X>** as the explanatory column.
- When the window appears choose **Analyze▶ Means, Anova/t-Test** from the popup option menu below the point plot. It turns out that the t-test for

two independent group means is a simplified version of what is called a "one-way ANOVA." ES7 covers this topic in Chapter 13.

The **Means for Oneway Anova** report shows the number of rows with values for Diastolic, each group's mean value, and the standard error s.

Means for Oneway Anova			
Level	**Number**	**Mean**	**Std Error**
FEMALES	13	71.0769	2.4265
MALES	16	77.3750	2.1872
Std Error uses a pooled estimate of error variance			

Note that the standard error can be different from what is in ES7 or in other computer output. JMP IN uses the methods for ANOVA (analysis of variance) to calculate error and degrees of freedom. The **t-Test** report automatically displays all the appropriate summary results, including the 95% confidence interval.

t-Test				
	Difference	**t-Test**	**DF**	**Prob>|t|**
Estimate	-6.2981	-1.928	27	0.0644
Std Error	3.2668			
Lower 95%	-13.0009			
Upper 95%	0.4048			
Assuming equal variances				

Summary Statistics: You can also do hypothesis testing and estimate the 95% confidence interval about the mean if all you have is summary statistics. For instance, in exercise 10.43 suppose you had only the table of sample sizes, means and standard deviations given in ES7. If these summary values are given with sufficient precision, you can recreate the **t-Test** report in the previous paragraph. Use the ONEWAY template documented on page 139 of the Appendix. For instance, if you type the summary statistics given in exercise 10.43 into ONEWAY the following table results.

EX10-043 using ONEWAY

7 Cols / 4 Rows	Group	N	Mean	SD	Freq	X	Y value
1	MALES	16	77.37	8.35	15	MALES	79.4575
2		•	•	•	1	MALES	46.0575
3	FEMALES	13	71.08	9.22	12	FEMALES	73.6371679

- Choose **Analyze>Fit Y by X** after the correct column rows have been specified as above.

- Choose **Analysis▸ Means, Anova/t-Test** from the option popup below the plot.

t-Test

	Difference	t-Test	DF	Prob>|t|
Estimate	-6.2900	-1.926	27	0.0647
Std Error	3.2662			
Lower 95%	-12.9917			
Upper 95%	0.4117			

Assuming equal variances

The only reason the values in the two **t-Test** reports did not match exactly is that the second report resulted from the values typed with only two decimal places of precision.

Exercise 10.57 (pg. 527)

The beginning steps for this exercise are the same as exercise 10.43. However, note from the point plot that there seems to be much more variability in Sample 2 than in Sample 1. The summary table given in exercise 10.57 in ES7 indicates that the standard deviation in Sample 2 is more than twice that in Sample 1 (17 versus 7).

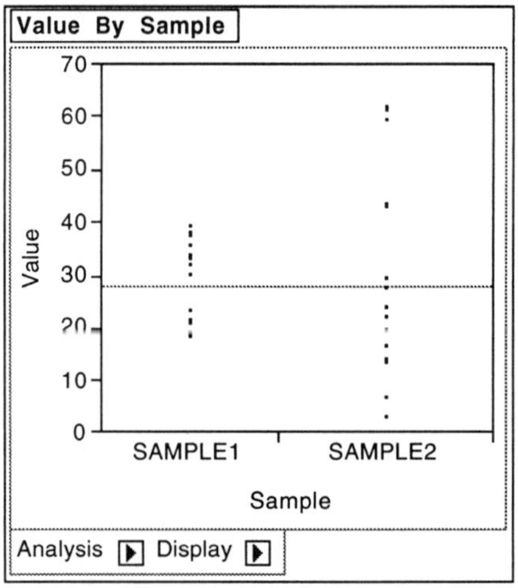

If you simply choose **Analysis▸ Means, Anova/t-Test** from the option popup below the plot the **t-Test** report is given below.

t-Test

	Difference	t-Test	DF	Prob>|t|
Estimate	2.8207	0.547	27	0.5888
Std Error	5.1557			
Lower 95%	-7.7578			
Upper 95%	13.3992			

Assuming equal variances

The "Assuming equal variances" note is there because the estimated standard error is calculated in a manner similar to formula 10-7 on page 513 of ES7. This formula is assuming that it makes sense to sum two similar value to get an "average" standard error.

If you choose **Analysis▸ Unequal Variances** from the option popup below the plot the **TESTS THAT THE VARIANCES ARE EQUAL** report gives a test at the bottom that allows the standard deviations to be different.

Tests that the Variances are Equal

Level	Count	Std Dev	MeanAbsDif to Mean	MeanAbsDif to Median
SAMPLE1	13	7.07297	6.13609	5.93846
SAMPLE2	16	17.41114	13.40078	13.05625

Test	F Ratio	DF Num	DF Den	Prob>F
O'Brien[.5]	4.6487	1	27	0.0401
Brown-Forsythe	4.3546	1	27	0.0465
Levene	5.7264	1	27	0.0239
Bartlett	8.7676	1	•	0.0031

Welch Anova testing Means Equal, allowing Std's Not Equal

F Ratio	DF Num	DF Den	Prob>F
0.3490	1	20.647	0.5611

t-Test
0.5908

There are four tests of equal variances. The O'Brien or Brown-Forsythe test are the most robust. In both cases, the p-value (Prob>F) is approximately 0.04, less than the α=.05 cut off most commonly used. This is a good indication that sample 2 has significantly more variability than does sample 1.

A t-test for testing the equality of two means with different variances is a special case of the more general F test shown at the bottom of the above report. The Welch Anova compares any number of means and does not assume that there is equal variability in the groups. In the example above, since there are just two groups, the t-test value is also given. The p-value is the same for both the t-test and the F test.

Exercise 10.63 (pg. 530)

Use the **RANNOR** template to generate normally distributed random variables. To simulate two variables, copy the formula of one and paste it into the JMP IN calculator for another new column.

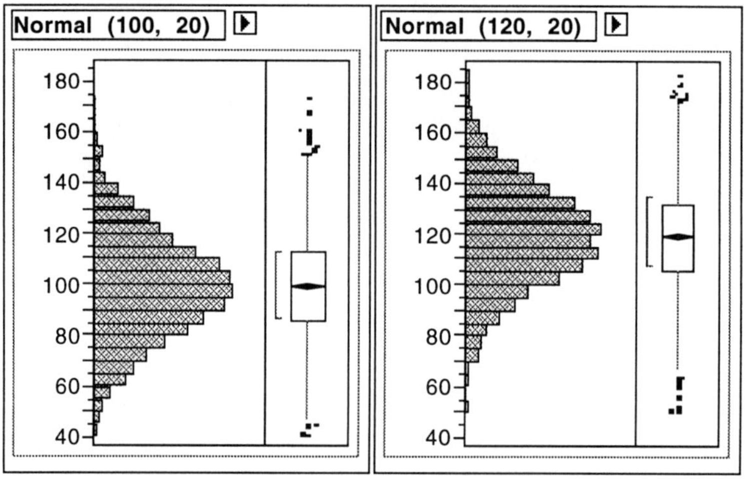

In part c again use RANNOR with # samples per trial = 8 and 800 rows. (You may want to save this data table for use with exercises 10.65-10.66.) Then use the **Tables>Group/Summary** command to calculate the means for each of the two normal variables separately for each group of samples within a trial. Then calculate the difference between the two means using the JMP IN calculator.

EX10-063 using RANNOR by Trial

5 Cols / By-Mode Off / 100 Rows	Trial	N Rows	Mean(Normal (100, 20))	Mean(Normal (120, 20))	Difference
1	1	8	101.5348	109.5765	8.041655
2	2	8	101.9955	122.467	20.47157
3	3	8	91.07145	110.508	19.4365
4	4	8	107.4373	121.3348	13.8975
5	5	8	97.41741	123.5438	26.12638
6	6	8	98.30296	112.9583	14.6553
7	7	8	103.8217	119.2962	15.47451

The **Distribution of Y** display shows that the average difference is approximately 20.5 with standard deviation 9.1.

Exercise 10.64 (pg. 531)

Exercise 10.64 is done exactly the same as exercise 10.63 except that the standard deviations of the two random normal columns are different.

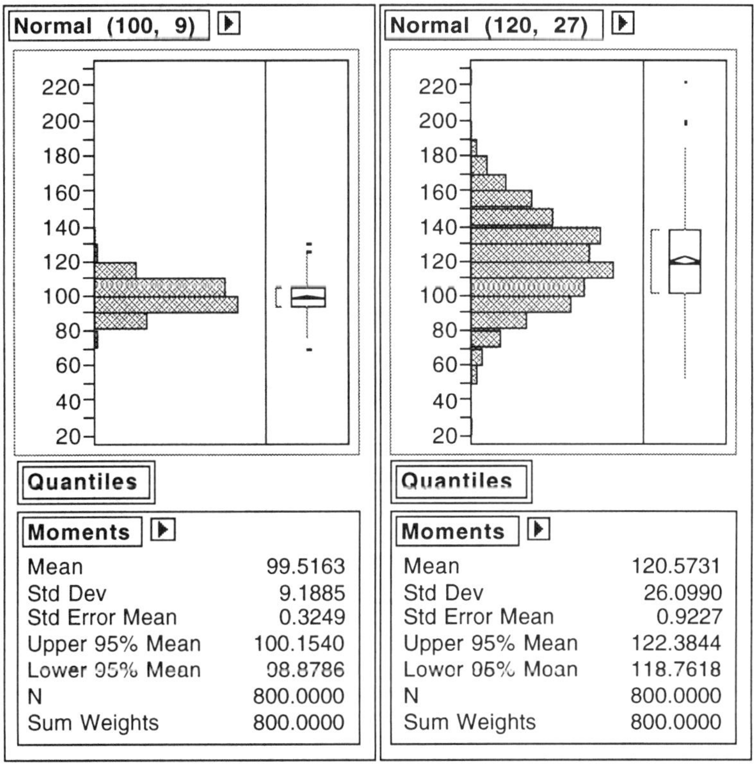

The Group/Summary table is calculated using the same method.

EX10-064 using RANNOR by Trial

5 Cols / By-Mode Off / 100 Rows	Trial	N Rows	Mean(Normal (100, 9))	Mean(Normal (120, 27))	Difference
1	1	8	97.6636	118.9438	21.28022
2	2	8	98.52659	101.9698	3.443245
3	3	8	95.63626	121.7025	26.0662
4	4	8	100.632	109.7475	9.115499
5	5	8	91.16867	126.9854	35.81671
6	6	8	101.9773	132.1516	30.17431
7	7	8	99.36463	119.0914	19.72673

The **Distribution of Y** display for the difference is shown below.

Difference

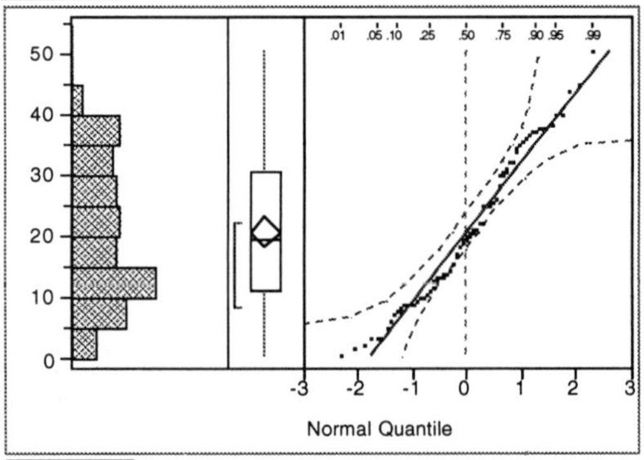

Quantiles

Moments

Mean	21.0568
Std Dev	11.4827
Std Error Mean	1.1483
Upper 95% Mean	23.3352
Lower 95% Mean	18.7784
N	100.0000
Sum Weights	100.0000

The mean and standard deviation above compare with the mean and standard error expected by the properties on pg. 512 of ES7.

Exercise 10.65 (pg. 531)

Exercise 10.65 is a repeat of exercise 10.63 with different sample sizes. Use the data table you built for exercise 10.63 change two things. 1) Specify # samples per trial = 20 and generate the required 100 • 20 rows. 2) Change the formula for the first randomly generated column so that it only calculates a sample value for 5 samples within each trial. The formula is as follows.

Mean⇐100
Standard Deviation⇐20
results $\begin{cases} \text{?normal}\bullet\textit{\textbf{Standard Deviation+Mean}}, & \text{if } \textit{Sample} \le 5 \\ \square, & \text{otherwise} \end{cases}$

A portion of the resulting table appears as follows.

EX10-065 using RANNOR

5 Cols / 2000 Rows	C # samples per trial	N Trial	N Sample	C Y Normal (100, 20)	C Y Normal (120, 20)
1	20	1	1	82.75	92.14
2	•	1	2	114.89	146.62
3	•	1	3	48.47	152.36
4	•	1	4	90.39	140.51
5	•	1	5	90.11	143.37
6	•	1	6	•	92.46
7	•	1	7	•	111.77

The remaining steps are the same as exercise 10.63.

Exercise 10.66 (pg. 531)

Use the formula in the RANEXP template to build a data table with the following structure.

EX10-66 using RANEXP

5 Cols / 1000 Rows	C # samples per trial	N Trial	N Sample	C Y Exp(50)	C Y Exp(80)
1	10	1	1	168.63	43.99
2	•	1	2	23.12	11.10
3	•	1	3	111.32	182.21
4	•	1	4	29.11	14.86
5	•	1	5	242.93	112.77
6	•	1	6	16.77	120.68
7	•	1	7	18.49	115.46

The remaining steps are the same as exercises 10.63-10.65.

Calculating values from the F distribution (pg. 536)

The FCDF template documented on page 155 of the Appendix calculates the left tail area under the curve for an F distribution. See exercise 10.83 below for an example of its use.

Exercise 10.83 (pg. 541-542)

The data for exercise 10.83 are on the student disk but the two groups of data are in separate columns. Use **Tables>Stack Columns** as in exercise 10.43 to restructure the data. The preferred method in JMP IN to test for unequal variances was shown in exercise 10.57 on page 102. In the Fit Y by X window, choose **Analysis▶ Unequal Variances** from the option popup below the plot and the **TESTS THAT THE VARIANCES ARE EQUAL** report appears.

Tests that the Variances are Equal

Level	Count	Std Dev	MeanAbsDif to Mean	MeanAbsDif to Median
MEN	16	7.886433	6.054688	5.812500
WOMEN	13	9.917609	7.976331	7.692308

Test	F Ratio	DF Num	DF Den	Prob>F
O'Brien[.5]	1.0108	1	27	0.3236
Brown-Forsythe	0.7748	1	27	0.3865
Levene	1.0222	1	27	0.3210
Bartlett	0.6804	1	•	0.4094

Welch Anova testing Means Equal, allowing Std's Not Equal

F Ratio	DF Num	DF Den	Prob>F
0.1460	1	22.704	0.7059

t-Test
0.3821

Since both O'Brien's test and the Brown-Forsythe test are not significant ($p > 0.3$), there is little evidence for unequal variability. If you wish to use the F test for the ratio of two variances, use the **FCDF** template.

EX10-083 using FCDF

4 Cols / 1 Rows	X	df-n	df-d	Pr(X≤x)
1	1.58	7	7	0.719947

You can use a JMP IN calculator formula to calculate the ratio of the two variances. Note also that the required p-value is 1 minus the value shown in PR(X ≤ X), or 0.280053.

Exercise 10.85 (pg. 542)

See the instructions for exercise 10.63 for the steps in using **RANNOR** to simulate two normal distributions. As in exercise 10.63, use Group/Summary to calculate statistics for each trial but calculate the standard deviation rather than the mean in this exercise.

EX10-085 using RANNOR by Trial

6 Cols / 100 Rows	Trial	N Rows	Std Dev(Normal (100, 20))	Std Dev(Normal (120, 20))	F star	critical?
1	1	8	16.44057	11.18209	2.16	0
2	2	8	12.45966	10.60082	1.38	0
3	3	8	11.94007	25.74509	4.65	1
4	4	8	19.52636	22.33345	1.31	0
5	5	8	21.31911	16.38374	1.69	0
6	6	8	21.11627	21.19693	1.01	0

The column **F star** is calculated as the ratio of the two variances, the larger over the smaller.

$$larger \Leftarrow \begin{cases} Std\ Dev(Normal\ (100,\ 20)), & \text{if } Std\ Dev(Normal\ (100,\ 20)) > Std\ Dev(Normal\ (120,\ 20)) \\ Std\ Dev(Normal\ (120,\ 20)), & \text{otherwise} \end{cases}$$

$$smaller \Leftarrow \begin{cases} Std\ Dev(Normal\ (120,\ 20)), & \text{if } Std\ Dev(Normal\ (100,\ 20)) > Std\ Dev(Normal\ (120,\ 20)) \\ Std\ Dev(Normal\ (100,\ 20)), & \text{otherwise} \end{cases}$$

$$\text{results } \left(\frac{larger}{smaller}\right)^2$$

Note that this calculation is different from that suggested in part c, which simply calculates the ratio of the first standard deviation over the second standard deviation. This method correspond to the two-tailed test shown in Illustration 10-15 beginning on pg. 538 of ES7. To count the number of trials whose observed F ratio exceeded the critical value, the following formula for Critical? was used.

$$F\ star > \text{F Quant}\left(0.975,\ \frac{7}{7}\text{ DF, centered at } 0\right)$$

The simulation results showed that 3 of 100 trials would have rejected the hypothesis of equal variance.

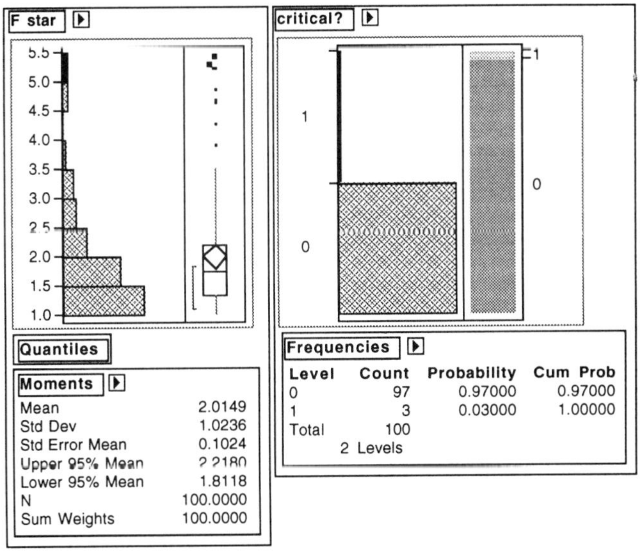

Quantiles

Moments

Mean	2.0149
Std Dev	1.0236
Std Error Mean	0.1024
Upper 95% Mean	2.2180
Lower 95% Mean	1.8118
N	100.0000
Sum Weights	100.0000

Frequencies

Level	Count	Probability	Cum Prob
0	97	0.97000	0.97000
1	3	0.03000	1.00000
Total	100		
	2 Levels		

You will get different results.

Exercise 10.86 (pg. 543)

As stated in the exercise, the steps are the same as in exercise 10.85 in every regard except that one of the variances is different.

Exercise 10.108 (pg. 554)

The methods for this exercise are similar to those described for exercises 10.63-10.66 and 10.85. Use the RANBIN template.

EX10-108 using RANBIN

8 Cols / 200 Rows	# samples per trial	Trial	Sample	Bin (100,.3)	Bin (100,.4)	p1'	p2'	p1'-p2'
1	200	1	1	31	39	0.31	0.39	-0.08
2	•	1	2	40	34	0.4	0.34	0.06
3	•	1	3	33	37	0.33	0.37	-0.04
4	•	1	4	27	39	0.27	0.39	-0.12
5	•	1	5	29	34	0.29	0.34	-0.05
6	•	1	6	28	36	0.28	0.36	-0.08
7	•	1	7	31	37	0.31	0.37	-0.06

Use the JMP IN calculator to calculate **p1'**, **p2'**, and the difference. The **Distribution of Y** display for the difference supports the notion that the column is normally distributed.

p1'-p2'

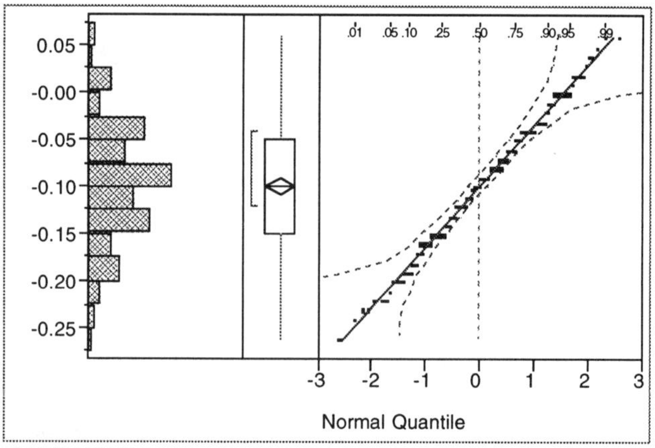

Quantiles

Moments

Mean	-0.0984
Std Dev	0.0641
Std Error Mean	0.0045
Upper 95% Mean	-0.0895
Lower 95% Mean	-0.1073
N	200.0000
Sum Weights	200.0000

Chapter 11
Additional Applications of Chi-Square

Categorical Responses

Response variables can be either continuous or categorical. The JMP categorical modeling types are nominal and ordinal, with nominal being the one most commonly used. In this chapter the methods for three situations are covered: 1) In a single sample, how do the observed responses compare with what you expected? 2) If you have two categorical variables, is one category related to the other or are the two categorical variables independent of one another. 3) The third situation is the same as 2) except that one of the categorical variables is fixed ahead of time rather than being determined in the outcome of the experiment. Also see chapters 9 and 10 in JSS for a discussion of categorical distributions and methods for fitting categorical models to crosstabulation tables.

Specific Applications in ES7

Calculating values for the chi-square distribution (pg. 581)

The use of the CHISQCDF template was described in chapter 9 on page 93. See the description there and the documentation in the Appendix on page 155. The chi-square value calculated in Illustration 11-2 is as below.

ILL11-2 using CHISQCDF

5 Cols / 1 Rows	df	X	Pr(X≤x)	Pr(X>x)
1	3	0.47	0.0746	0.9254

A chi-square value this small is very likely.

Exercise 11.9 (pg. 582)

There is no need, with JMP IN, to calculate the observed and expected values by hand. Simply create a data table with the observed values and frequency counts and use the **Distribution of Y** platform. The data table for exercise 11.9 is as follows.

EX11-09

2 Cols / 4 Rows	Rating	Count
1	1 excellent	44
2	2 good	145
3	3 fair	133
4	4 poor	28

The Rating column is a character column with ordinal modeling type. The default ordering of a character column is alphabetic but you do not want the order to be: excellent, fair, good, poor. By adding a numeric prefix, the sorted order will be correct. Count is numeric with a Frequency modeling type.

The **Distribution of Y** display for Rating with Count as a Freq variable shows the observed proportions. Choose the ▶ **Test probabilities** option from the popup above the histogram. Enter the hypothesized probabilities in the spaces provided and click **Done** when complete.

Test Probabilities

Level	Estim Prob	Hypoth Prob
1 excellent	0.12571	0.16000
2 good	0.41429	0.38000
3 fair	0.38000	0.41000
4 poor	0.08000	0.05000

Click then Enter Hypothesized Probabilities.
Omitted values will use estimated probabilities.
The remaining values will be scaled to sum to 1 minus omitted

Done Help

The final tables and figures describe the observed and tested proportions.

Frequencies ▶

Level	Count	Probability	Cum Prob
1 excellent	44	0.12571	0.12571
2 good	145	0.41429	0.54000
3 fair	133	0.38000	0.92000
4 poor	28	0.08000	1.00000
Total	350		

4 Levels

Test Probabilities

Level	Estim Prob	Hypoth Prob
1 excellent	0.12571	0.16000
2 good	0.41429	0.38000
3 fair	0.38000	0.41000
4 poor	0.08000	0.05000

Test	Chi-square	DF	Prob>Chisq
Likelihood Ratio	9.9372	3	0.0191
Pearson	10.7224	3	0.0133

The Pearson chi-square value given above is the same as that calculated by formula 11-1.

Chi-square for contingency table (pg. 596)

The most common difficulty in the analysis of contingency tables arises from the improper layout of the data table. By "improper" we mean that there are ways to enter the correct *values* into a JMP data table whose *structure* will not allow appropriate analyses. For example, one data table structure that will work is shown below. It uses the values in Table 11-10 on page 592 of ES7.

TAB11-10

3 Cols / 6 Rows	Residence	Count	Proposal
1	Urban	143	Favor
2	Urban	57	Oppose
3	Suburban	98	Favor
4	Suburban	102	Oppose
5	Rural	13	Favor
6	Rural	87	Oppose

Structure, modeling type, and roles Notice that there is one row per cell of the contingency table. One of the columns identifies the row of the contingency table, in this case Residence. One column identifies the column of the contingency table, in this case Proposal. And, since frequency counts are given in the contingency table, it is convenient to have a COUNT column to act in the Frequency role. If you are testing independence, mark one column as X and the other Y; it does not matter which. If you are testing for homogeneity (one of the columns is fixed), mark the column that is not free to vary with the X role and the column whose response is random with Y.

If your data has already been entered into a table as follows, all is not lost. Just use the **Tables>Stack Columns** command to restructure the table as above.

TAB11-10 in need of Stack Columns

3 Cols / 3 Rows	Residence	Favor	Oppose
1	Urban	143	57
2	Suburban	98	102
3	Rural	13	87

Once the data has been properly arranged, use **Analyze>Fit Y by X** to perform the contingency table analysis. The following roles were specified: Residence>**X>**, Proposal>**Y>**, and Count>**Freq>**. The

Crosstabs▶ popup was used to show the expected counts and cell chi-square values.

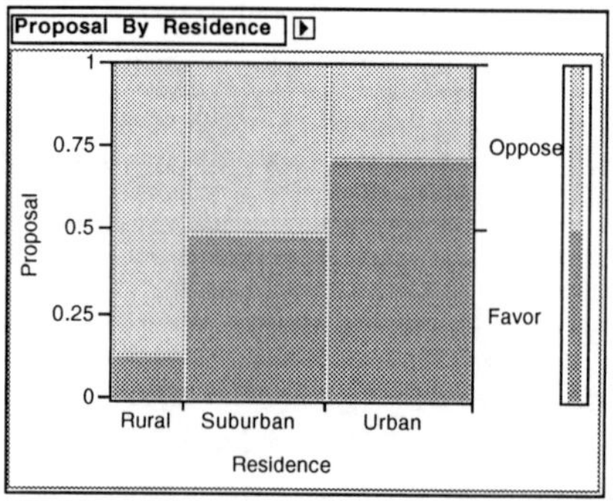

The proportion favoring increases from its lowest rate in rural areas to high rates in urban residences.

Crosstabs

Proposal	Count Expected Cell Chi^2	Rural	Suburban	Urban	
		Residence			
	Favor	13 50.8 28.1268	98 101.6 0.1276	143 101.6 16.8697	254
	Oppose	87 49.2 29.0415	102 98.4 0.1317	57 98.4 17.4183	246
		100	200	200	500

Tests

Source	DF	-LogLikelihood	RSquare (U)
Model	2	49.75871	0.1436
Error	497	296.75087	
C Total	499	346.50959	
Total Count	500		

Test	ChiSquare	Prob>ChiSq
Likelihood Ratio	99.517	<.0001
Pearson	91.715	<.0001

Chapter 12
Analysis of Variance

One-way Anova in Fit Y by X

In JMP IN there is no distinction between the comparison of the means of two independent samples as in Chapter 10 and a single-factor ANOVA as described in this chapter. The methods used in JMP IN to compare two independent groups are the same as the methods used to compare more than two independent groups. In JMP IN the t-Test comparison of two groups is handled as a special case of the Anova comparison of multiple groups. See chapter 6 in JSS for a discussion of differences between two means and chapter 7 for the more general methods for comparing many means.

Specific Applications in ES7

ANOVA (pg. 626-627)

Illustration 12-4 uses the data in Table 12-7 on pg. 623 of ES7. The TAB12-7 file contains the data.

TAB12-7

3 Cols / 6 Rows	RIGHTEYE	LEFTEYE	BOTHEYES
1	12	10	16
2	10	17	14
3	18	16	16
4	12	13	11
5	14	•	20
6	•	•	21

This data table is not structured in a way that JMP IN can use the information to compare the groups. However, it is easy to rearrange the table. Use the methods in exercise 10.43 in chapter 10 page 99 to restructure the data table using the **Tables>Stack Columns** command.

TAB12-7 Stacked

2 Cols / 18 Rows	Accuracy	Method
1	12	RIGHTEYE
2	10	LEFTEYE
3	16	BOTHEYES
4	10	RIGHTEYE
5	17	LEFTEYE
6	14	BOTHEYES
7	18	RIGHTEYE
8	16	LEFTEYE
9	16	BOTHEYES
10	12	RIGHTEYE
11	13	LEFTEYE
12	11	BOTHEYES
13	14	RIGHTEYE
14	•	LEFTEYE
15	20	BOTHEYES
16	•	RIGHTEYE
17	•	LEFTEYE
18	21	BOTHEYES

Note that since there are an unequal number of observations in each group, some of the values of ACCURACY are missing. You do not have to do anything to "fix" this; JMP IN automatically ignores missing values and calculates the appropriate results. With ACCURACY marked as a continuous response column and Method marked as a nominal explanatory column, the Fit Y by X platform displays the point plot showing the distribution of values in each **Method** group. Choose the **Analysis ▶ Means, Anova/t-Test** option in the popup below the plot to perform the oneway Anova.

Accuracy By Method

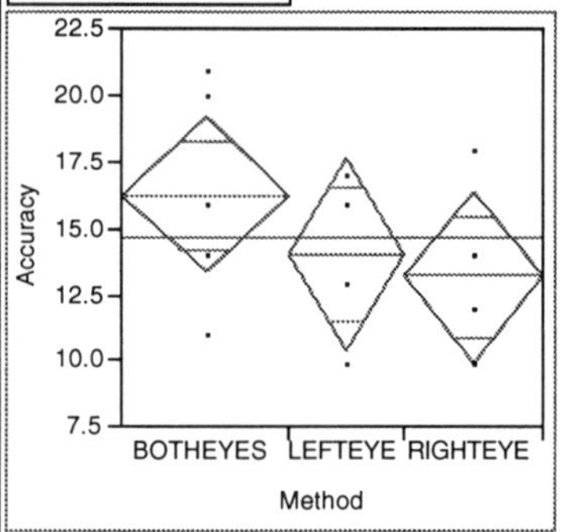

Analysis ▶ Display ▶

Oneway Anova ▶

Summary of Fit

RSquare	0.176613
RSquare Adj	0.039382
Root Mean Square Error	3.368151
Mean of Response	14.66667
Observations (or Sum Wgts)	15

Analysis of Variance

Source	DF	Sum of Squares	Mean Square	F Ratio
Model	2	29.20000	14.6000	1.2870
Error	12	136.13333	11.3444	Prob>F
C Total	14	165.33333	11.8095	0.3116

Means for Oneway Anova

Level	Number	Mean	Std Error
BOTHEYES	6	16.3333	1.3750
LEFTEYE	4	14.0000	1.6841
RIGHTEYE	5	13.2000	1.5063

Std Error uses a pooled estimate of error variance

Notice that the unequal number of observations in each case have been taken care of. That is, even though there are 18 rows in the data table, there are only 15 rows that have non-missing values for both **Accuracy** and **Method** so only these rows are included in the analysis. The point plot has also been modified to include diamonds centered on the group mean and whose height represent the 95% confidence interval about each of the group means. Recall that if there are repeated values for the response variable within a group, you may want to choose the **Display ▶ Jitter** option in the popup below the plot to make all of the points visible.

Exercise 12.19 (pg. 631-632)

The data in EX12-19 are properly structured for JMP IN but you must change the modeling type of HS_GROUP to Nominal. A portion of the data table with the correct modeling types and roles is shown below.

EX12-19

2 Cols / 25 Rows	C Y KSW_SCOR	N X HS_GROUP
1	16	1
2	22	1
3	15	1
4	20	1
5	23	1
6	16	1
7	21	1
8	21	2

The **Analysis of Variance** report from the Fit Y by X platform is the same as that shown on pg. 632 of ES7.

Analysis of Variance

Source	DF	Sum of Squares	Mean Square	F Ratio
Model	2	214.74390	107.372	4.6509
Error	22	507.89610	23.086	**Prob>F**
C Total	24	722.64000	30.110	0.0207

Exercises 12.23, 12.24 (pg. 634-635)

The 95% confidence intervals about group means are automatically displayed by means diamonds in JMP IN. They automatically appear when you choose the **Analysis▶ Means, Anova/t-Test** option in the popup below the plot to perform the oneway analysis of variance.

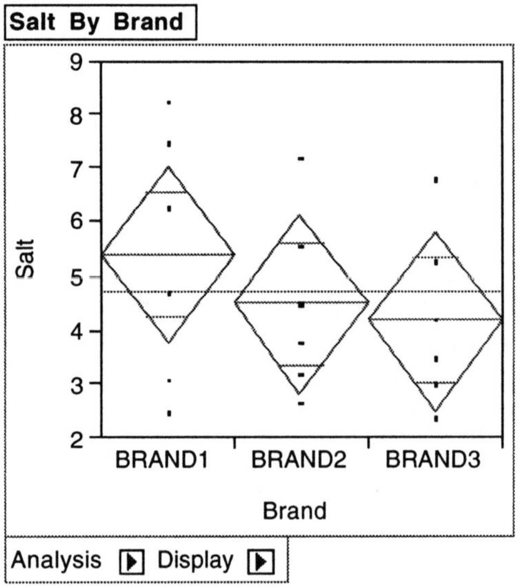

The default confidence level displayed is 95%. If you want a different level, choose an **Analysis▸ Set alpha level** option.

Chapter 13
Linear Correlation and Regression Analysis

Scatter plots in Fit Y by X

You saw in chapter 3 how to look at bivariate data when both variables were quantitative. JMP IN always displays the scatter plot first and then allows your analysis to unfold from that point. This often involves estimating the linear correlation between the continuous columns or fitting a line or curve through the points. See JSS chapter 8 for additional discussion of fitting curves through points and regression.

Specific Applications in ES7

Exercise 13.5, finding the extensions and SS's using the calculator (pg. 653-654)

The FORM3-2 template calculates all of the required values and its use was discussed in Chapter 3. See **Chapter 3: Specific Applications in ES7: Finding the correlation coefficient** on page 55. The template is documented on page 133.

EX13-05 using FORM3-2

6 Cols / 6 Rows	X	Y	X^2	Y^2	X•Y	r
1	45	22	2025	484	990	0.94
2	52	26	2704	676	1352	0.94
3	49	21	2401	441	1029	0.94
4	60	28	3600	784	1680	0.94
5	67	33	4489	1089	2211	0.94
6	61	32	3721	1024	1952	0.94

The simplest way to obtain the SS(x), SS(y), and SS(xy) terms is to use the JMP IN calculator. Bring up the formula display for the calculated column r and click on each of the intermediate calculations.

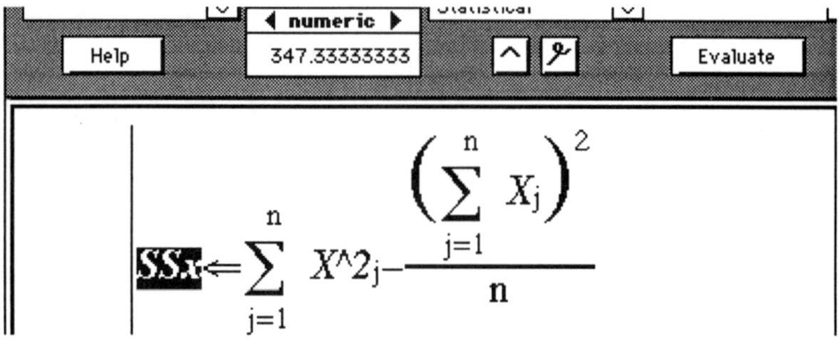

When you click on a term in the formula display window, it highlights by displaying white letters on a black background. The calculator also displays the calculated value of the highlighted term in the "constant box." In the figure above, you can see that the numeric constant value of ***SSx*** is approximately 347.33.

Note: If you accidentally change the formula by clicking on something and then typing on the keyboard, just choose **Edit>Undo** to put thing back the way they were. You can also close the data table without saving and start over.

Exercise 13.7 (pg. 655)

Correlations and methods for obtaining their estimate were discussed in Chapter 3. See **Chapter 3: Specific Applications in ES7: Finding the correlation coefficient** on page 55.

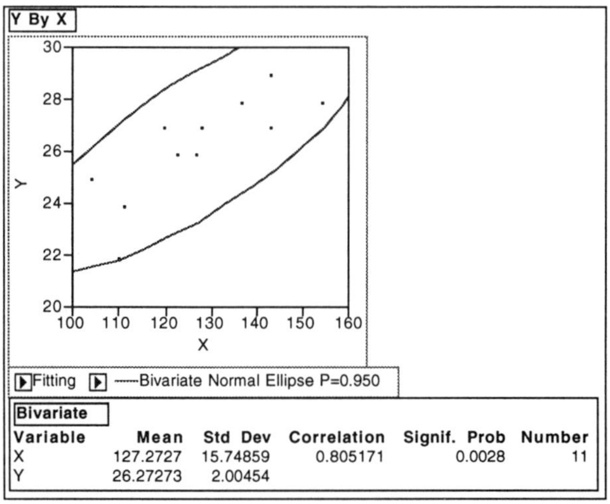

The narrower the ellipse, the higher the correlation. A circle means zero correlation.

Exercises 13.31 (pg. 669), 13.43 (pg. 675)

The data in exercise 13.31 are on the student disk. Both columns are numeric and continuous. Mark the predictor column with the X role and the response column with Y the role.

EX13-31

2 Cols / 10 Rows	C X X	C Y Y
1	12	15
2	14	25
3	16	30
4	20	30
5	23	30
6	46	80
7	50	90
8	48	95
9	50	110
10	55	130

- Choose **Analyze>Fit Y by X**.
- Choose the **Fitting ▶ Fit line** option. The equation of the line of best fit and all related statistics are automatically displayed.

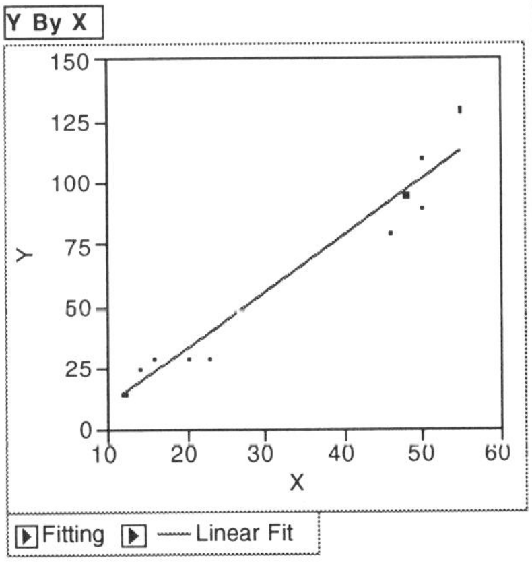

Linear Fit

Y = -13.414 + 2.3028 X

Summary of Fit

RSquare	0.947433
RSquare Adj	0.940863
Root Mean Square Error	10.17383
Mean of Response	63.5
Observations (or Sum Wgts)	10

Analysis of Variance

Source	DF	Sum of Squares	Mean Square	F Ratio
Model	1	14924.446	14924.4	144.1882
Error	8	828.054	103.5	Prob>F
C Total	9	15752.500		<.0001

Parameter Estimates

Term	Estimate	Std Error	t Ratio	Prob>|t|
Intercept	-13.41352	7.167862	-1.87	0.0982
X	2.3027999	0.191775	12.01	<.0001

The steps for exercise 13.43 are identical to those for exercise 13.31.

Confid. and Pred. interval bands (pg. 683-684)

Confidence bands and prediction bands can be added to fitting any curve. In exercise 13.31 above the best fitting line was displayed. Notice that this added an additional popup menu below the plot labeled Linear Fit.

- Choose **Linear Fit ▶ Confid curves: Fit** to obtain confidence bands for the mean value of y.
- Choose **Linear Fit ▶ Confid curves: Indiv** to obtain prediction bands for the individual values of y.

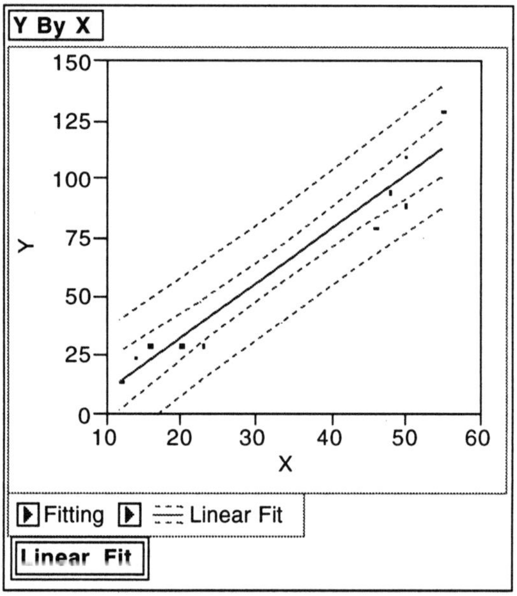

Chapter 14 Elements of Nonparametric Statistics

Using the nonparametric options of Fit Y by X

The Mann-Whitney U test is one of the nonparametric test options in the oneway layout of Fit Y by X platform. Recall from chapter 10 (Inferences Involving Two Populations) and chapter 12 (Analysis of Variance) that the parametric way to compare two groups is with the independent samples t-Test. Or, more generally, the oneway Anova. The way this is done in JMP IN is with the Fit Y by X platform. What you did in those chapters is mark the qualitative column that identifies the groups as a ominal [X] explanatory column. The quantitative variable that was the measured response was marked as a [C]ontinuous [Y] column. To do nonparametric comparisons you do exactly the same thing. Yes, even though you might think that you should mark the Y column as ordinal, don't do it. A portion of the data table for Illustration 14-6 is thus:

ILL14-06

2 Cols / 20 Rows	[N] [X] Source	[C] [Y] Exam score
9	A	49
10	A	78
11	B	72
12	B	62
13	B	91
14	B	88
15	B	90
16	B	74

As ES7 describes, many nonparametric methods are based on ranks. Illustration 14-6 shows exam grades on two different exams and their ranks. These ranks are one possible way to score the exam grades. Ranking the exam grades is what is called the *Wilcoxon score*. This is why the way you do the

Mann-Whitney U test is to choose the **Analysis▶ Nonpar-Wilcoxon** option from the popup below the point plot in the Fit Y by X window.

Wilcoxon / Kruskal-Wallis Tests (Rank Sums)

Level	Count	Score Sum	Score Mean	(Mean-Mean0)/Std0
A	10	86	8.6000	-1.401
B	10	124	12.4000	1.401

2-Sample Test, Normal Approximation

S	Z	Prob>|Z|
124	1.40110	0.1612

1-way Test, Chi-Square Approximation

ChiSquare	DF	Prob>ChiSq
2.0706	1	0.1502

Just as oneway Anova is a generalization of the two group t-Test, the Wilcoxon Kruskal-Wallis test is a generalization of the two group Mann-Whitney U test. Another name for this test is the Wilcoxon rank sum test. Sorry about the jargon. I find it helpful to remember to begin the analysis like you are going to do an ANOVA but don't; choose one of the nonparametric analysis options.

Note that the sum of the ranks for each group is labeled **Score Sum** in the above report. For Illustration 14-6 these rank sums correspond to R_a and R_b in formula 14-4 and 14-3, respectively. The hypothesis of no group difference is calculated with a z statistic as in formula 14-7. The z value is shown above in the 2-Sample Test, Normal Approximation report.

Sign rank test in Distribution of Y

Single-sample hypothesis test procedure

JMP IN does not calculate the sign test but it does do a signed rank test. The signed rank test assumes the distributions are symmetric while the sign test does not. In ES7 section 8.4 you saw how to find the confidence interval for a mean and test whether the mean was equal to a specified value, assuming normality. See **Chapter 8: Finding the confidence interval for mean: Case 1: Data values** on page 81. The method to use when variability is estimated from the data was discussed in ES7 section 9.1. See **Chapter 9: Finding the confidence interval for mean: Case 1: Data values** on page 86. In both cases you used **Analyze>Distribution of Y** and choose ▶ **Test mean=value** from the popup at the top of the histogram. As in both of these cases, to do a nonparametric test, actually of whether the *median* values is equal to a specified number, enter a value for **Specify Hypothesized Value for**

Mean: but before you click **OK**, check the **If you want a nonparametric test, check here: Wilcoxon Signed-Rank Test** optional check box.

In exercise 14.3 December temperatures were recorded for 20 randomly selected days. The exercise asks to test whether the median is 48 degrees. The **Test** [*sic.* Median] **=value** report is shown below.

Test Mean=value		
Hypothesized Value		48
Actual Estimate		44.3
	t Test	**Signed-Rank**
Test Statistic	-2.0174	-53.500
Prob > |t|	0.0580	0.030
Prob > t	0.9710	0.985
Prob < t	0.0290	0.015

Two-sample hypothesis test procedure

JMP IN does the signed rank test to compare groups. See the discussion of **Using the nonparametric options of Fit Y by X** in this chapter on page 123.

Using the nonparametric correlations option of Correlation of Y's

The **Analyze>Correlation of Y's** command automatically displays a table of the parametric (Pearson) correlation matrix of all of the Y variables. To obtain the Spearman rank correlations, choose ☑ **Correlations-Nonpar** from the popup the lower-left border of the Correlation of Y's window. It automatically displays Spearman's rho and Kendall's tau. To add other nonparametric measures or to remove Kendall's tau, check or uncheck the options in the **Nonparametric measures of association** ▶ check list. For Illustration 14-11 the final Correlation of Y's reports appear below.

Correlations

Variable	Judge 1	Judge 2	Judge 3	Judge 4
Judge 1	1.0000	-1.0000	1.0000	-0.2000
Judge 2	-1.0000	1.0000	-1.0000	0.2000
Judge 3	1.0000	-1.0000	1.0000	-0.2000
Judge 4	-0.2000	0.2000	-0.2000	1.0000

Nonparametric Measures of Association ▶

Variable	by Variable	Spearman Rho	Prob>|Rho|
Judge 2	Judge 1	-1.0000	0.0000
Judge 3	Judge 1	1.0000	0.0000
Judge 3	Judge 2	-1.0000	0.0000
Judge 4	Judge 1	-0.2000	0.7471
Judge 4	Judge 2	0.2000	0.7471
Judge 4	Judge 3	-0.2000	0.7471

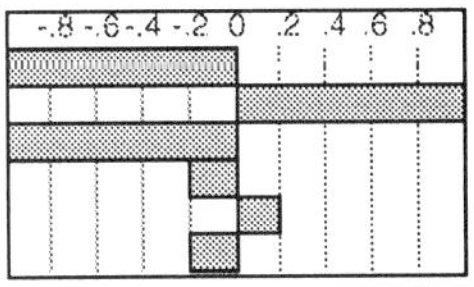

Specific Applications in ES7

Finding the Mann-Whitney statistic, Exercise 14.23 (pg. 719)

The columns in the original data file for exercise 14.23 must be stacked as in performing a oneway analysis of variance or independent group t-Test.

- Mark **Gender** as nominal and **Pulse** as continuous.
- Choose **Analyze>Fit Y by X**.
- Specify **Pulse >Y>** and **Gender >X>**. A point plot appears.
- Choose the **Analysis ▶ Nonpar-Wilcoxon** option from the popup below the plot.
- For descriptive purposes, it is a good idea to also choose the **Analysis ▶ Quantiles** option from the popup below the plot.

Pulse By Gender

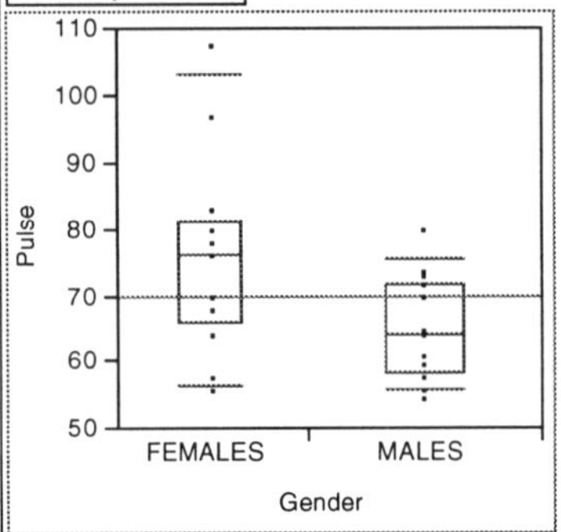

Analysis ▶ Display ▶

Quantiles

Level	minimum	10.0%	25.0%	median	75.0%	90.0%	maximum
FEMALES	56	56.8	66	76	81.5	103.6	108
MALES	55	55.7	58.5	64.5	72	75.8	80

Wilcoxon / Kruskal-Wallis Tests (Rank Sums)

Level	Count	Score Sum	Score Mean	(Mean-Mean0)/Std0
FEMALES	13	243	18.6923	2.087
MALES	16	192	12.0000	-2.087

2-Sample Test, Normal Approximation

S	Z	Prob>|Z|
243	2.08739	0.0369

1-way Test, Chi-Square Approximation

ChiSquare	DF	Prob>ChiSq
4.4494	1	0.0349

The median pulse rate for females of 76 seems significantly higher than the median male rate of 64.5 by the rank sum test ($p < .0369$).

The runs statistic (pg. 723)

The runs test is not supported by JMP IN.

Exercise 14.43 (pg. 732)

In exercise 14.43 the ranks of the two columns are calculated and then a Pearson's correlation is obtained on the ranks. The Spearman rank correlation can also be obtained this way but there is a simpler way.

- Choose **Analyze>Correlation of Y's**.
- Specify AGE **>Add>** and MINCON **>Add>**. A correlation matrix appears.
- Choose **Correlations-Nonpar** from the popup the lower-left border of the Correlation of Y's window.

It automatically displays Spearman's rho and Kendall's tau. To remove Kendall's tau, uncheck the option in the **Nonparametric measures of association** ▶ check list. The results for exercise 14.43 are shown below.

Correlations

Variable	AGE	MINCON
AGE	1.0000	0.6884
MINCON	0.6884	1.0000

Nonparametric Measures of Association ▶

Variable	by Variable	Spearman Rho	Prob>|Rho|
MINCON	AGE	0.7531	0.0047

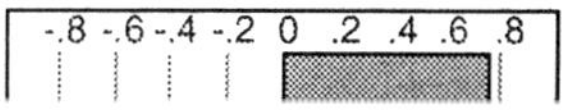

Calculating ranks The easiest way to calculate the ranks for columns is to let the **Distribution of Y** platform do it for you. Using the data for exercise 14.43 the steps are as follows.

- Choose **Analyze>Distribution of Y**.
- Specify AGE **>Add>** and MINCON **>Add>**. A correlation matrix appears.
- Choose **Save ranks averaged** from the popup the lower-left border of the **Distribution of Y** window.

This adds two new columns to the data table.

EX14-43

4 Cols / 12 Rows	AGE	MINCON	RankAvgd AGE	RankAvgd MINCON
1	82	170	11	12
2	83	40	12	8
3	64	64	8	11
4	53	5	6	2.5
5	47	15	4	6
6	50	5	5	2.5
7	70	48	9	9
8	62	34	7	7
9	34	3	3	1
10	27	7	1	4
11	75	50	10	10
12	28	10	2	5

The Pearson correlation between the ranks is indeed the Spearman rank correlation.

Bivariate

Variable	Mean	Std Dev	Correlation	Signif. Prob	Number
RankAvgd AGE	6.5	3.605551	0.753066	0.0047	12
RankAvgd MINCON	6.5	3.599242			

Appendix

Data Analysis Templates

JMP and JMP IN do not do everything you can imagine a statistics or graphics program might do. However, using the JMP calculator you can create templates that do a wide variety of things. Templates are empty JMP data tables with calculator formulas embedded in columns. Templates are *not* SAS transport files; they are JMP files. SAS transport files can not have formulas; they are just data. JMP files can contain both data and formulas that automatically recalculate when the data changes. To use JMP templates, choose **File>Open**. Do not attempt **File>Import**; it will not work. This Appendix lists the templates used in ES7.

The templates are not yet available on the ES7 student data disk. They may available in both Macintosh and Windows format from your instructor or on your school's file server. Or, they can be downloaded from the web. If you have web access, point your browser to:
http://views.vcu.edu/~albest/JMP.html

Or download them using anonymous FTP using the following settings.
FTP server: ruby.VCU.edu
Username: anonymous
Password: anonymous
Directory: /best /ES7

In addition, the templates are documented here so that you can build them yourself.

DOTPLOT

The DOTPLOT template has two columns:

DotPlot

2 Cols / 0 Rows	sorted values	Frequency

The first column contains data values while the second column, **Frequency**, has the formula:

$$\begin{cases} 1, & \text{if } sorted\ values \neq sorted\ values_{i-1} \\ (Frequency_{i-1})+1, & \text{otherwise} \end{cases}$$

Use of DotPlot

- Sort the original data table by the column of interest: Choose **Table>Sort** and fill in the sort dialog.
- Copy the values of the column of interest into the clipboard: Highlight the column, and choose **Edit>Copy**.
- Choose **File>Open** and select the DOTPLOT file.
- Paste the values on the clipboard into the first column: Change the column name from **sorted values** to the original column name from the source dataset. Highlight the first column. Choose **Edit>Paste**.
- Display the Dot plot: Choose **Analyze>Fit Y by X**. Specify Frequency as the Y column and the first column as X.

FORM2-6

Formula 2-6 calculates the sample variance as the mean squared deviation. These calculations can be performed using the FORM2-6 template. It has 6 columns.

FORM2-6

6 Cols / 0 Rows	X	X bar	deviation	deviation^2	s - squared	s

All of the columns are calculated except the first. The JMP IN calculator formulas for the last 5 columns are as follows.

X bar uses the Statistical>Mean function:

$$\overline{X}$$

deviation uses the subtraction button:

$$X - X\ bar$$

deviation^2 uses the squared button:

$$deviation^2$$

s - squared uses the Statistical>Sum function, the divide button, the left parenthesis button, and the subtraction button. The Terms>N function calculates the sample size:

$$\frac{\sum_{j=1}^{n} \textit{deviation}^\wedge 2_j}{(n-1)}$$

s uses the square root button:

$$\sqrt{s\text{ - }squared}$$

Of course, you could simply use the Statistical>Std Deviation function to do all the calculations in a single operation.

Use of FORM2-6

- Copy the values of the column of interest into the clipboard: Highlight the column, and choose **Edit>Copy**.
- Choose **File>Open** and select the FORM2-6 file.
- Paste the values on the clipboard into the first column: Change the column name from X to the original column name from the source dataset. Highlight the first column. Choose **Edit>Paste**.

The calculations automatically occur and the results appear in the data table.

FORM2-10

Formula 2-10 calculates the sample variance using a shortcut formula. These calculations can be performed using the FORM2-10 template. It has 4 columns.

FORM2-10

4 Cols / 0 Rows	X	X^2	s - squared	s

All of the columns are calculated except the first. The JMP IN calculator formulas for the last 3 columns are as follows.

X^2 uses the squared button:

$$X2$$

s - squared uses the Statistical>Sum function, the divide button, the left parenthesis button, and the subtraction button. The Terms>N function calculates the sample size.:

$$\frac{\left(\sum_{j=1}^{n} X\text{^}2_j - \left(\frac{\left(\sum_{j=1}^{n} X_j\right)^2}{n}\right)\right)}{(n-1)}$$

s uses the square root button:

$$\sqrt{s\text{ - }squared}$$

Use of FORM2-10

- Copy the values of the column of interest into the clipboard: Highlight the column, and choose **Edit>Copy**.
- Choose **File>Open** and select the FORM2-10 file.
- Paste the values on the clipboard into the first column: Change the column name from X to the original column name from the source dataset. Highlight the first column. Choose **Edit>Paste**.

The calculations automatically occur and the results appear in the data table.

NORMRULE

On pgs. 104-107 in ES7 an empirical rule is given for checking normality. To check a set of values with this rule, use the NORMRULE template. The template as one data column, SCORE, and three calculated columns.

NORMRULE

4 Cols / 0 Rows	C Y SCORE	N Y ~68%	N Y ~95%	N Y ~99.7%

The formula for the column ~95% is as follows (and the others are similar).

$$mean \Leftarrow \overline{SCORE}$$
$$SD \Leftarrow \text{std } SCORE$$
$$\text{results } (mean - SD \cdot 2) \leq SCORE \leq (mean + SD \cdot 2)$$

The formulas calculate 0/1 values according to whether the original variable is within 1, 2, or 3 standard deviations.

Use of NORMRULE

- Copy the values of the column of interest into the clipboard: Highlight the column, and choose **Edit>Copy**.
- Choose **File>Open** and select the NORMRULE file.
- Paste the values on the clipboard into the first column: Change the column name from SCORE to the original column name from the source dataset. Highlight the first column. Choose **Edit>Paste**.

  The calculations automatically occur and the results appear in the data table. To calculate the observed proportions, calculate the mean of these calculated columns.

- Choose **Analyze>Distribution of Y**.
- Select the columns **~68%, ~95%, ~99.7%** (using Shift-click). The **Distribution of Y** report displays the proportion of rows with values of 1 and you can compare these with the expected percentage.

FORM3-2

This template calculates the correlation coefficient as in ES7 Table 3-11 on pg. 153. It uses formulas 3-2 through 3-4.

6 Cols / 0 Rows	X	Y	X^2	Y^2	X•Y	r

The template has two data columns and four calculated columns.

X^2 uses the square button, as does **Y^2**:

$$X^2$$

X•Y uses the multiply button:

$$X \bullet Y$$

r is where most of the work is done:

$$SSx \Leftarrow \sum_{j=1}^{n} X\text{^}2_j - \frac{\left(\sum_{j=1}^{n} X_j\right)^2}{n}$$

$$SSy \Leftarrow \sum_{j=1}^{n} Y\text{^}2_j - \frac{\left(\sum_{j=1}^{n} Y_j\right)^2}{n}$$

$$SSxy \Leftarrow \sum_{j=1}^{n} X \bullet Y_j - \frac{\left(\sum_{j=1}^{n} X_j \bullet \sum_{j=1}^{n} Y_j\right)}{n}$$

$$\text{results } \frac{SSxy}{\sqrt{SSx \bullet SSy}}$$

Use of FORM3-2

- Copy the values of the columns of interest into the clipboard: Highlight the columns, and choose **Edit>Copy**.
- Choose **File>Open** and select the FORM3-2 file.
- Paste the values on the clipboard into the first two columns: Highlight the first two columns using shift-click. Choose **Edit>Paste**.

The calculations automatically occur and the results appear in the data table.

FORM8-1

Formula 8-1 on pg. 366 of ES7 calculates the confidence interval using four pieces of information: The sample size is typed into the column N. The point estimate is Mean (X). The known σ is typed in Std Dev (X). Degree of confidence is specific by alpha (α) ; use .05 for this column if you want a 95% confidence interval. These four values are entered into the first four columns.

FORM8-1

7 Cols / 1 Rows	Sigma	n	X bar	alpha (a)	SE Mean	Lower CI	Upper CI
1	1.76	100	10.22	0.05	0.176	9.875046	10.56495

The formula for **SE Mean** is:

$$\frac{Std\ Dev\ (X)}{\sqrt{n}}$$

The formula for **Lower CI** is:

$$Mean\ (X) - \text{normQuant}\left(1-\frac{alpha\ (a)}{2}\right) \bullet SE\ Mean$$

The formula for **Upper CI** is similar.

Use of FORM8-1

- Type the values for the first four columns. The formulas automatically recalculate.

FORM8-3

Formula 8-3 on page 391 of ES7 calculates a test statistic z★, given an observed mean from n samples where the hypothesized mean is μ and σ are known. The four observed or known values are typed into the first four columns.

FORM8-3

8 Cols / 1 Rows	Mu	Sigma	n	X Bar	z star	Prob |z|	Prob > z	Prob < z
1	4.5	2.87	40	3.975	-1.157	0.247301	0.87635	0.12365

The formula for **z star** is:

$$\frac{X\ Bar - Mu_1}{\left(\frac{Sigma_1}{\sqrt{n}}\right)}$$

The formula for each of the p-value calculations are as follows:

Prob |z|	**Prob > z**	**Prob < z**
normDist(-|*z star*|)•2	1–normDist(*z star*)	normDist(*z star*)

Use of FORM8-3

- Type the values for the first four columns. If you have multiple samples for the same hypothesized values, you need only type in **Mu** and **Sigma** in the first row. The values for **Mu** and **Sigma** in the first row used with all values of **n** and **X Bar**. When any of the values in the first 4 columns change, the last 4 calculated columns change.

FORM9-1

The formula to make a confidence interval estimation using an observed mean, standard deviation, and sample size is given in formula 9-1 on page 439 of ES7. In this template, the sample size is typed into the column **N**. The point estimate is **Mean (X)**. The sample standard deviation is typed in **Std Dev (X)**. Degree of confidence is specific by **alpha (α)**; use .05 for this column if you want a 95% confidence interval. These four values are entered into the first four columns.

FORM9-1

7 Cols / 1 Rows	n	Mean (X)	Std Dev (X)	alpha (a)	SE Mean	Lower CI	Upper CI
1	•	•	•	0.05	•	•	•

The formula for **SE Mean** is:

$$\frac{Std\ Dev\ (X)}{\sqrt{n}}$$

The formula for **Lower CI** is:

$$Mean\ (X) - \text{t Quant}\left(1-\frac{alpha\ (a)}{2}, n-1 \text{ DF, centered at } 0\right) \bullet SE\ Mean$$

The formula for **Upper CI** is similar.

Use of FORM9-1

- Type the values for the first four columns. The formulas automatically recalculate.

FORM9-2

Formula 9-2 on page 441 of ES7 calculates a test statistic t★, given an observed mean and standard deviation from n samples where the hypothesized mean is μ is known. The four observed or known values are typed into the first four columns.

FORM9-2

8 Cols / 1 Rows	Mu	n	X Bar	Std Dev	t star	Prob |t|	Prob > t	Prob < t
1	4.9	22	5.1	1.17	0.802	0.43166	0.21583	0.78417

The formula for **t star** is:

$$\frac{X\,Bar - Mu_1}{\left(\frac{Std\,Dev_1}{\sqrt{n}}\right)}$$

The formula for each of the p-value calculators are as follows:

Prob |t|

$$\text{t Dist}(-|t\ star|, n-1\ \text{DF, centered at } 0)\bullet 2$$

Prob > t

$$1-\text{t Dist}(t\ star, n-1\ \text{DF, centered at } 0)$$

Prob < t

$$\text{t Dist}(t\ star, n-1\ \text{DF, centered at } 0)$$

Use of FORM9-2

- Type the values for the first four columns. If you have multiple samples for the same hypothesized values, you need only type in Mu in the first row. The values for Mu and in the first row are used with all values of n and X Bar, and Std Dev. When any of the values in the first 4 columns change, the last 4 calculated values change.

FORM9-6

Formula 9-6 on pg. 455 of ES7 calculates the confidence interval for p', the observed binomial proportion. The formula uses, alpha to specify the (1-alpha)% confidence, the sample size n, and the observed number of successes, x.

FORM9-6

7 Cols / 1 Rows	n	x	alpha (a)	Obs:p'	Std Err	Lower CI	Upper CI
1	200	17	0.1	0.085	0.020	0.053	0.117

The formula for **Obs:p'**, the observed proportion is:

$$\frac{x}{n}$$

The Std Err is calculated using the observed proportion:

$$\sqrt{\frac{Obs{:}p' \bullet (1 - Obs{:}p')}{n}}$$

The confidence intervals are calculated with JMP IN formulas as given in FORM8-1 shown on page 134.

Use of FORM9-6

- Type the values for the first three columns. The formulas automatically recalculate.

FORM9-9

Formula 9-9 on pg. 459 of ES7 calculates the test statistic z★ for testing hypotheses concerning the binomial parameter p, given an a sample of size n and an observed number of successes, x. The three observed or hypothesized values are typed into the first three columns.

FORM9-9

9 Cols / 1 Rows	Ho:p	n	x	Obs:p'	z star	Prob |z|	Prob > z	Prob < z	Messages
1	0.61	350	235	0.6714	2.356	0.018464	0.009232	0.990768	

The formula for **Obs:p'**, the observed proportion is:

$$\frac{x}{n}$$

The formula for **z star** is as given in formula 9-9:

$$\frac{Obs{:}p' - Ho{:}p_1}{\sqrt{\dfrac{Ho{:}p \bullet (1 - Ho{:}p)}{n}}}$$

The formulas for **Prob |z|**, **Prob > z**, and **Prob <z** are identical to the formulas for the corresponding columns in FORM8-3.

Use of FORM9-9

- Type the values for the first three columns. If you have multiple samples for the same hypothesized value, you need only type in Ho:p in the first row. The value for Ho:p in the first row is used with all values of n and X. When any of the values in the first 3 columns change, the calculated values change.

  Note that there is a **Messages** column. Notes appears in this column if any of the following are not true: $n > 20$, $np > 5$, $nq > 5$, or if p is not a valid proportion.

FORM9-10

Formula 9-10 on page 474 of ES7 calculates an observed $\chi^{2\star}$ value from a hypothesized value of the population standard deviation, and an observed value of a sample standard deviation s derived from a sample of size n.

FORM9-10

6 Cols / 1 Rows	Ho:Sigma	n	Obs:s	chi^2 star	Prob > Chi^2	Prob ≤ Chi^2
1	10	12	14	21.560	0.028012	0.971988

The formula for **chi^2 star** is:

$$\frac{(n-1) \bullet Obs{:}s^2}{Ho{:}Sigma_1^2}$$

The formula for **Prob>Chi2** is:

1–chi-square Dist(*chi^2 star* , *n* –1 DF, centered at 0)

The formula for **Prob≤Chi^2** is as above but without the "1-".

Use of FORM9-10

- Type the values for the first three columns. Note that the values to be entered are standard deviations, not variances. Note also that only the first row's value of **Ho: Sigma** is used in the calculations. This allows you to simulate a number of sample distributions and compare them all with a single hypothesized value.

ONEWAY

The **ONEWAY** JMP template makes use of the fact that all you need to know to compare the means of groups are the "sufficient statistics." These statistics are the sample size, mean, and standard deviation (or variance) for each group. Given just these three pieces of data for each group, the **ONEWAY** template synthesizes values so that JMP IN thinks you have all the observations in each group. An example comparison of two groups from exercise 10.43 is shown below.

ONEWAY

7 Cols / 4 Rows	Group	N	Mean	SD	Freq	X	Y value
1	FEMALES	13	71.0769231	9.22371525	12	FEMALES	73.6351214
2		•	•	•	1	FEMALES	40.3785431
3	MALES	16	77.375	8.34965069	15	MALES	79.4624127
4		•	•	•	1	MALES	46.0638099

The first 4 columns of information identify the groups and provide the necessary summary statistics. The last three columns are calculated.

The formula for **Freq** calculates **Freq**=1 for the even rows and **Freq**= N_{Group}-1 for the odd rows for every other row.

$$\begin{cases} 1, & \text{if i mod 2=0} \\ N-1, & \text{otherwise} \end{cases}$$

The argument for the if clause is true for the even numbered rows and false for the odd numbered rows.

The **X** formula duplicates the values entered into **Group** for every other row.

$$\begin{cases} Group_{i-1}, & \text{if i mod 2=0} \\ Group, & \text{otherwise} \end{cases}$$

The formula for **Y value** calculates a value above and below the mean so that when the N-1 values of the odd rows are averaged in with the single values in the even rows, the correct group means and standard deviation result.

$$\begin{cases} N_{i-1} \bullet Mean_{i-1} - (N_{i-1} - 1) \bullet Y\ value_{i-1}, & \text{if i mod 2=0} \\ Mean\ , & \text{if } SD \leq 0 \\ Mean + \sqrt{\dfrac{SD^2}{N}}, & \text{otherwise} \end{cases}$$

It also handles the case of a single observation in a group.

Use of ONEWAY

The data table needs twice as many rows as groups. If you have K groups, you enter the information into 2 • K rows. Type or paste the values for the first four columns and the last three columns automatically recalculate. Then use the Fit Y by X platform to perform the t-Test or oneway analysis of variance.

- Open the ONEWAY template and use **Rows>Add Rows** to result in a data table with twice as many rows as groups. For a two group comparison you need 4 rows.
- Type the values for Group, N, Mean and SD into only the ***odd*** numbered rows. Verify that the calculated value for X is the same as Group and that the calculated values for Freq are 1 and N-1 for each group.
- Mark Freq with the Freq role, X with the X role, and Y value with the Y role.
- Choose **Analyze>Fit Y by X**.
- Choose the **Analysis▸ Means, Anova/t-Test** option from the popup below the plot.
- Verify that the n's and means in the **Means for Oneway Anova** report are what you keyed in.

Means for Oneway Anova

Level	Number	Mean	Std Error
FEMALES	13	71.0769	2.4265
MALES	16	77.3750	2.1872

Std Error uses a pooled estimate of error variance

The statistics to compare the groups are in the **Analysis of Variance** report, or if there are only two groups, in the **t-Test** report.

Random Number Generation Templates

General Use

The following templates were designed for the general case where you want to generate random numbers from a particular distribution. The instructions for each template shows you how to specify the sampling distribution.

In each template, you must also tell JMP IN how many cases to generate. This is done as follows. The templates generate T=# trials for K=# samples per trial. There are two situations: 1) You want to generate K values from a population (in a single trial). Or, 2) you want to generate K values T times (for multiple trials). In both cases you type the value of K into the data table. The difference is in the number of rows in the final JMP data table. The two cases are described below.

Single trial, K samples Augment the table so that it has K rows. Since the template comes with 1 row, add K-1 more. For example if you want 10 samples, choose **Rows>Add rows** and specify **# Rows to Add: 9**.

T trials, K samples Augment the table so that it has T•K rows. Since the template comes with 1 row, add (T•K)-1 more. For example if you want 5 trials, each with 10 samples, you need a total of 50 rows. Choose **Rows>Add Rows** and specify **# Rows to Add: 49**.

General steps In each case you open the template, change the parameter(s) that describe the sampling distribution, specify K (the number of samples per trial), and add sufficient rows to generate the complete sample. Complete instructions are given for RANNOR, below. For the other random number templates, only the differences between the RANNOR case and the other populations are described.

RANNOR

The RANNOR JMP template generates # samples per trial=K, random normal values with mean and standard deviation specified in a formula.

Use of RANNOR

- Choose **File>Open** and select RANNOR.

RANNOR

4 Cols / 1 Rows	# samples per trial	Trial	Sample	Normal (M, S)
1	10	1	•	-0.07992

- Double-click above the column name Normal (M, S).

  The Column Information dialog appears. Change the column name if you like.

- Click the mini-formula button in the lower left corner to display the formula in the JMP calculator.

  Normal (M, S) uses variables to specify the mean and standard deviation. These parameter values appear in the top of the JMP formula:

  Mean⇐0
  Standard Deviation⇐1
  results ?normal•***Standard Deviation+Mean***

- Change the values of the mean and standard deviation used.
  - Click on the zero value to the right of ***Mean***⇐.
  - Type the number you want. (It appears in the constant box.)
  - Press the Return or Enter key.
  - Click on the one to the right of ***Standard Deviation***⇐.
  - Type the number you want.
  - Press the Return or Enter key.
  - Verify that the formula is displaying the correct mean and SD.
  - Close the calculator window.
- Change the value of # samples per trial in the data table to the number you require.
- Determine the total number of rows required (see page 142) and add the appropriate number of rows
  - Choose **Rows>Add Rows**.
  - Specify **# Rows to Add:** (see page 142).

  The additional rows are added and the random numbers generated. For example, the final result for K=5 samples in T=2 trials for Mean=100, and SD=10 is given below.

RANNOR

4 Cols / 10 Rows	C # samples per trial	N Trial	N Sample	C Normal (M, S)
1	5	1	1	103.9658
2	•	1	2	89.16682
3	•	1	3	122.3829
4	•	1	4	93.75768
5	•	1	5	105.1366
6	•	2	1	99.13391
7	•	2	2	94.05821
8	•	2	3	100.3189
9	•	2	4	92.62201
10	•	2	5	97.49861

RANCOIN

The **RANCOIN** template generates K=**# samples per trial** of random Bernoulli values. Bernoulli trials have values of 0 or 1. If you want the values to be "Heads" and "Tails" or "Tack up" and "Tack side", add another formula column using **Condition>IF** to recode the 0/1 values. If all you need to do is calculate T•K cases, each with .5 probability, then all you need to specify is the value for **# samples per trial** and then generate the appropriate number of rows.

RANCOIN

4 Cols / 1 Rows	C # samples per trial	N Trial	N Sample	C Coin 0/1
1	10	1	•	0

The formula for the column **Coin 0/1** is as follows.

?binomial(1, probability 0.5)

Use of RANCOIN

- See the instructions for **RANNOR** to see how to open the template, change column names, and display the calculator formula.
- Change the values of the probability in the formula for **Coin 0/1**, if required.
- Specify the number of samples per trial and generate the appropriate number of rows.

RANBIN

The **RANBIN** JMP template generates K = **# samples per trial** of random binomial values using parameters: N = # Bernoulli experiments, each with probability of success = p. The result is the number of successes in N experiments.

RANBIN

4 Cols / 1 Rows	# samples per trial (C)	Trial (N)	Sample (N)	Binomial (n,p) (C)
1	10	1	1	6

The formula for the column Binomial (n,p) is as follows.

$$\begin{array}{l} N \Leftarrow 10 \\ p \Leftarrow 0.5 \\ \text{results ?binomial}(N, \text{ probability } p) \end{array}$$

Use of RANBIN

- See the instructions for RANNOR to see how to open the template, change column names, and display the calculator formula.
- Change the values of parameters N and/or p, in the formula for Binomial (n,p).
- Specify the number of samples per trial and generate the appropriate number of rows.

RANUNI

The RANUNI JMP template generates K=# samples per trial of uniformly distributed random values between a minimum and maximum value. The default range is between min=0 and max=1. If all you need to do is calculate T•K cases, each within the range (0,1), then all you need to specify is the value for # samples per trial and then generate the appropriate number of rows.

RANUNI

4 Cols / 1 Rows	# samples per trial (C)	Trial (N)	Sample (N)	Uniform (L, H) (C)
1	10	1	1	0.325199

The formula for the column Uniform (L, H) is as follows.

$$\begin{array}{l} \mathit{Minimum} \Leftarrow 0 \\ \mathit{Maximum} \Leftarrow 1 \\ \text{results } \mathit{Minimum} + \text{?uniform} \bullet (\mathit{Maximum} - \mathit{Minimum}) \end{array}$$

Use of RANUNI

- See the instructions for RANNOR to see how to open the template, change column names, and display the calculator formula.
- Change the values of the parameters ***Minimum*** and ***Maximum***, in the formula for Uniform (L, H).
- Specify the number of samples per trial and generate the appropriate number of rows.

RANDIE6

The RANDIE6 template rolls a single 6-sided die. It generates the integers 1, 2, 3, 4, 5, and 6, all equally likely. If all you need to do is roll the die K times, generate a table with that many rows.

RANDIE6

4 Cols / 1 Rows	# samples per trial	Trial	Sample	6 sided die
1	10	1	1	5

The formula for the column 6 Sided Die is as follows.

$$\lfloor ?\text{uniform} \bullet 6+1 \rfloor$$

The formula uses the Numeric>Floor function, as well as the Random> Uniform function.

Use of RANDIE6

- See the instructions for RANNOR to see how to open the template, change column names, and display the calculator formula.
- Specify the number of samples per trial and generate the appropriate number of rows.

RANINT

The RANINT template generates random integers. It is very similar to RANDIE6 in this way except that RANDIE6 only generates values between 1 and 6. RANINT generates whole numbers between a minimum and maximum value (inclusive).

RANINT

4 Cols / 1 Rows	C ▢ # samples per trial	N ▢ Trial	N ▢ Sample	O ▢ digits [L,H]
1	10	1	1	7

The formula for the column digits (L,H) is as follows.

$$
\begin{array}{|l}
\boldsymbol{Minimum} \Leftarrow 0 \\
\boldsymbol{Maximum} \Leftarrow 9 \\
\boldsymbol{range} \Leftarrow \boldsymbol{Maximum} - \boldsymbol{Minimum} + 1 \\
\text{results } \lfloor ?\text{uniform} \bullet (\boldsymbol{range}) + 1 \rfloor + (\boldsymbol{Minimum} - 1)
\end{array}
$$

Use of RANINT

- See the instructions for RANNOR to see how to open the template, change column names, and display the calculator formula.
- Change the parameter values of the ***Minimum*** and/or ***Maximum*** in the formula for digits (L,H).
- Specify the number of samples per trial and generate the appropriate number of rows.

RANEXP

The RANEXP JMP template generates random values from the single parameter exponential distribution. The parameter specifies both the mean and standard deviation of the distribution.

RANEXP

4 Cols / 1 Rows	C ▢ # samples per trial	N ▢ Trial	N ▢ Sample	C ▢ Exponential (M)
1	10	1	1	0.35508

The formula for the column Exponential(M) is as follows.

$$
\begin{array}{|l}
\boldsymbol{Mean\ \&\ SD} \Leftarrow 1 \\
\text{results } \boldsymbol{Mean\ \&\ SD} \bullet ?\text{exponential}
\end{array}
$$

Use of RANEXP

- See the instructions for RANNOR.
- Change the values of the parameter in the formula for Exponential(M).
- Specify the number of samples per trial and generate the appropriate number of rows.

RANGAM

The RANGAM JMP template generates K=# samples per trial of from a gamma distribution with shape and scale parameters.

RANGAM

4 Cols / 1 Rows	C # samples per trial	N Trial	N Sample	C Gamma(A,B)
1	10	1	•	0.856217

The formula for the column Gamma(A, B) is as follows.

Shape (A)⇐1
Scale (B)⇐1
results ?gamma(alpha ***Shape (A)***)•***Scale (B)***

Use of RANGAM

- See the instructions for RANNOR to see how to open the template, change column names, and display the calculator formula.
- Change the values of the ***Shape (A)*** and ***Scale (B)***, if required. If 2• ***Shape*** is integer and ***Scale*** =2, then a chi-square with df=2•***Shape*** is generated.
- Specify the number of samples per trial and generate the appropriate number of rows.

RANDISC

This JMP template generates integers with specific probabilities. That is, for the integers 1, 2, 3, ... the template generates random values with probability Prob_1, Prob_2, Prob_3, ...

RANDISC

5 Cols / 10 Rows	C Prob	C # samples per trial	N Trial	N Sample	N Discrete
1	0.2	10	1	1	1
2	0.3	•	1	2	2
3	0.3	•	1	3	1
4	0.1	•	1	4	1
5	0.1	•	1	5	2
6	•	•	1	6	2
7	•	•	1	7	3
8	•	•	1	8	2
9	•	•	1	9	3
10	•	•	1	10	5

In addition to columns similar to those of the other random number generation JMP templates, an additional column has been added to specify the probabilities needed. Note that in the figure above, there are 5 probabilities in the Prob column. Thus, the integers from 1 to 5 are generated. If you want 5 discrete values simulated, enter only 5 values into the first 5 rows of Prob; all other values <u>must</u> be missing. It is also up to you to make sure that these values sum to 1.

The last column, Discrete, has a rather complicated looking formula.

$$
\left|\begin{array}{l}
\boldsymbol{p} \Leftarrow 0\\
\boldsymbol{u} \Leftarrow ?\text{uniform}\\
\text{results} \displaystyle\sum_{j=1}^{\text{N}\,Prob} \left|\begin{array}{l} \boldsymbol{p\_low} \Leftarrow \boldsymbol{p}\\ \boldsymbol{p} \Leftarrow \boldsymbol{p} + Prob_{\text{j}}\\ \text{results } \boldsymbol{p\_low} < \boldsymbol{u} \end{array}\right.
\end{array}\right.
$$

Do not change anything in this formula. All it does is generate a uniformly distributed random number (0, 1) and then, for the number of non-missing values of Prob, it maps the probabilities into integer values.

Use of RANDISC

- See the instructions for RANNOR to see how to open the template. Do not change any of the formulas in the template.
- Change the values of the first column. Key in the probabilities associated with each of the integers in the first I columns. If you want 5 integers, key in values in the first 5 rows.
- Specify the number of samples per trial and generate the appropriate number of rows.

To generate another set of values, type over any of the values in Prob. When values in Prob change (even when they change to the same value), JMP recalculates another set of random values for Discrete.

RANCHISQ

The RANCHISQ template generates values from a chi-square distribution with specified degrees of freedom.

RANCHISQ

4 Cols / 1 Rows	C # samples per trial	N Trial	N Sample	C Chi-square(df)
1	10	1	1	7.744491

The formula for **Chi-square(df)** is as follows:

$$\begin{array}{|l} df \Leftarrow 1 \\ \text{results ?gamma(alpha } \boldsymbol{df}) \end{array}$$

Use of RANCHISQ

- Change the value of the ***df*** parameter in the formula. Note that ***df*** should be an integer.
- Specify the number of samples per trial and generate the appropriate number of rows.

Distribution Calculators

BINOMIAL

The BINOMIAL template calculates the binomial probabilities (pdf) and cumulative binomial probabilities (cdf) for x = 0 , 1, 2, …*n* number of successes each with probability of success *p*.

When you first open the data table, it has only 1 row.

BINOMIAL

5 Cols / 1 Rows	C p	C x	C nCx	C Pr (X = x)	C Pr(X ≤ x)
1	0.4	0	1	1	1

The formula for **x** is:

$$i-1$$

The nCX column calculates number of combinations of "n choose x."

$$\left| \begin{array}{l} n \Leftarrow n-1 \\ \text{results} \displaystyle\prod_{j=1}^{n} \left(\frac{j}{\begin{cases} j, & \text{if } j \le x \\ 1, & \text{otherwise} \end{cases} \bullet \begin{cases} j, & \text{if } j \le (n-x) \\ 1, & \text{otherwise} \end{cases}} \right) \end{array} \right.$$

The pdf is calculated in **Pr(X = x)** using the following formula:

$$\left| \begin{array}{l} p \Leftarrow p_1 \\ q \Leftarrow 1-p \\ n \Leftarrow n-1 \\ \#ways \Leftarrow nCx \\ \text{results } \#ways \bullet p^{x} \bullet (q)^{(n-x)} \end{array} \right.$$

The cdf is calculated in **Pr(X ≤ x)** using the following formula:

$$\sum_{j=1}^{i} Pr\,(X = x)_j$$

Use of BINOMIAL

To use the binomial calculator there are two steps: Type in the value of p and specify the appropriate number of rows in the data table. The steps for calculating the pdf and cdf for p=.5 and n=5 are given below.

- Choose **File>Open** and specify file BINOMIAL.
- Type .5 into the first row of P.

  To specify the value of n, you must end up with **n+1** rows. Since the template comes with 1 row, add n more.

- Choose **Rows>Add Rows** and specify **# Rows to add: 5**. The resulting table is shown below.

BINOMIAL (5, 0.5)

5 Cols / 6 Rows	p	x	nCx	Pr (X = x)	Pr(X ≤ x)
1	0.5	0	1	0.03125	0.03125
2	•	1	5	0.15625	0.1875
3	•	2	10	0.3125	0.5
4	•	3	10	0.3125	0.8125
5	•	4	5	0.15625	0.96875
6	•	5	1	0.03125	1

Note that X ranges from 0 to n and that there are n+1 rows. All of the columns are calculated except the first. You need not change any of the formulas.

PROBNORM

The JMP template PROBNORM calculates the normal pdf from a normal distribution with specified mean and standard deviation. The first column is the observed value and the second column is a calculated probability.

PROBNORM

2 Cols / 1 Rows	obser.	Norm pdf
1	0	0.3989

The formula for Norm pdf is as follows:

$$\begin{array}{|l} \boldsymbol{Mean} \Leftarrow 0 \\ \boldsymbol{SD} \Leftarrow 1 \\ \text{results normDensity}\left(\frac{(obser.-\boldsymbol{Mean})}{\boldsymbol{SD}}\right) \end{array}$$

Use of PROBNORM

To use the template, you must first specify the mean and standard deviation of the known normal distribution. The values are changed in the formula for Norm pdf.

- Specify the mean and standard deviation of the normal distribution as follows.
  - Click on the zero value to the right of ***Mean***⇐.
  - Type the number you want. (It appears in the constant box.)
  - Press the Return or Enter key.
  - Click on the one to the right of ***SD***⇐.

- Type the number you want.
- Press the **Return** or **Enter** key.
- Verify that the formula is displaying the correct mean and SD.
- Close the calculator window.

- Type the observed value(s) into the first column and the calculated pdf results.

NORMCDF

The NORMCDF template converts observed values into z scores for a normal distribution with specified mean and standard deviation. The normal cdf of the z score is also calculated. In JMP the normal cdf is the area under the curve below the observed value. That is, the cdf is Pr(Z <z).

The template has three columns for data and two calculated columns.

NORMCDF

7 Cols / 1 Rows	Mean	Std Dev	X	z	Pr(Z≤z)	Pr(Z>z)	pdf
1	0	1	0	0.00	0.5	0.5	0.3909

The values in the first row for the population Mean and Std Dev are used to calculate the z score for all of the observed values in X, the observed normal score.

The formula for **z** is the standard normal score:

$$\frac{X - Mean_1}{Std\ Dev_1}$$

The formula for **Pr(Z ≤ z)** is the left-tail area under the curve:

$$\text{normDist}(z)$$

The formula for **Pr(Z ≥ z)** is the right-tail area under the curve:

$$1 - Pr(Z \leq z)$$

The formula for **pdf** is the probability density function:

$$\text{normDensity}(z)$$

Note: the function used above is not visible in the JMP IN calculator function browser. The only place you can get it is in this template.

Use of NORMCDF

There are two things to specify when using the **NORMCDF** template, the parameters of the normal distribution and the observed scores.

- Type specified values for **Mean** and **Std Dev** into the first row. Values used in other rows are *not* used.

  There are generally two situations for values of the observed score. The two situation are: You have a single score and you have multiple scores. If you have a single score, do the following.

- Type the observed value into **X** and the calculated cdf results.

  If you have a number of values, simply add more rows and type in more values or use the calculator to generate other values.

  If you have multiple scores, create a table with as many rows as you need by using **Rows>Add rows**. For **X** values either 1) type in values, 2) paste values in from another data table, or 3) generate values using the JMP calculator.

TCDF

The **TCDF** template calculates the cumulative probability from -∞ to a specified value of t for the Student's t-distribution with **df** degrees of freedom. The first two columns contain numeric values. The last column is the calculated cdf.

TCDF

5 Cols / 1 Rows	df	t	Pr(T≤t)	Pr(T>t)	pdf
1	1	0.00	0.5	0.5	0.3183

The formula for **Pr(T ≤ t)** is the left-tail area under the curve:

$$\begin{cases} \text{t Dist}(t,\ df\ \text{DF, centered at } 0), & \text{if } t \neq \bullet \text{ and } df > 0 \\ \square, & \text{otherwise} \end{cases}$$

The formula for **Pr(T> t)** is the right-tail area under the curve.

The formula for **pdf** the probability density function.

$$\frac{\text{t Dist}(t+\Delta,\ df_1\ \text{DF, centered at } 0) - \text{t Dist}(t-\Delta,\ df_1\ \text{DF, centered at } 0)}{2 \bullet \Delta}$$

Note: the formula used above is not easy to obtain in the JMP IN calculator. The best place to get it is in this template.

Use of TCDF

- Type specified values for t and df into the first row.

TDIST

The TDIST template calculates the value of t whose cumulative probability is p, for a Student's t-distribution with df degrees of freedom. The first two columns contain numeric values. The last column is the calculated t.

TDIST

3 Cols / 1 Rows	Pr(T≤t)	df	t
1	0.05	1	-6.3138

The formula for **t**, is:

$$\begin{cases} \text{t Quant}(Pr(T\leq t),\ df\ \text{DF, centered at 0}), & \text{if } Pr(T\leq t)\neq\bullet \text{ and } df>0 \\ \square & \text{otherwise} \end{cases}$$

Use of TDIST

- Type specified values for Pr(T ≤ t) and df into the first row.

CHISQCDF

The **CHISQCDF** template calculates the cumulative distribution function for values of the chi-square distribution. The first two columns contain numeric values. The last column is the calculated cdf.

CHISQCDF

5 Cols / 1 Rows	df	X	Pr(X≤x)	Pr(X>x)	pdf
1	27	47.25	0.9907	0.0093	0.0024

The formula for **Pr(X≤x)** is the left-tail area under the curve.

$$\begin{cases} \text{chi-square Dist}(X,\ df\ \text{DF, centered at 0}), & \text{if } X\geq 0 \text{ and } df>0 \\ \square & \text{otherwise} \end{cases}$$

The formula for **Pr(X> x)** is the right-tail area under the curve.

The formula for **pdf** is the probability density function.

$$\frac{\text{chi-square Dist}(X+\Delta,\ df_1\ \text{DF, centered at 0}) - \text{chi-square Dist}(X-\Delta,\ df_1\ \text{DF, centered at 0})}{2\bullet\Delta}$$

Note: the formula used above is not easy to obtain in the JMP IN calculator. The best place to get it is in this template.

Use of CHISQCDF

- Type specified values for **X**, the observed value of chi-square, and **DF** into the first row. The area under the curve to the left of the value calculates.

FCDF

The **FCDF** JMP template calculates the cumulative probability of a value from the F distribution with the specified degrees of freedom.

FCDF

6 Cols / 1 Rows	f	df-n	df-d	Pr(F≤f)	Pr(F>f)	pdf
1	1.00	1	1	0.5	0.5	0.1592

The formula for **Pr(F ≤ f)** is as follows.

$$\begin{cases} \text{F Dist}\left(f, \dfrac{df\text{-}n}{df\text{-}d}\ \text{DF, centered at } 0\right), & \text{if } f \geq 0 \text{ and } df\text{-}n > 0 \text{ and } df\text{-}n > 0 \\ \square & \text{otherwise} \end{cases}$$

The formula for **Pr(F > f)** is the right-tail area under the curve.

The formula for **pdf** is the probability density function.

$$\frac{\text{F Dist}\left(f+\Delta, \dfrac{df\text{-}n_1}{df\text{-}d_1}\ \text{DF, centered at } 0\right) - \text{F Dist}\left(f-\Delta, \dfrac{df\text{-}n_1}{df\text{-}d_1}\ \text{DF, centered at } 0\right)}{2 \bullet \Delta}$$

Note: the formula used above is not easy to obtain in the JMP IN calculator. The best place to get it is in this template.

Use of FCDF

- Type specified values for **f**, the observed value of F distribution, and the numerator and denominator **df-n** and **df-d** into the first row. The area under the curve calculates.

Index